T0338218

BREAKTHROUGH PROJECT PORTFOLIO MANAGEMENT

Achieving the Next Level
of Capability and Optimization

MURALI KULATHUMANI, MBA, CSM

Library of Congress Cataloging-in-Publication Data
Names: Kulathumani, Murali, 1974– author.
Title: Breakthrough project portfolio management : achieving the next level
  of capability and optimization / Murali Kulathumani.
Description: Plantation, FL : J. Ross Publishing, [2017] | Includes index.
Identifiers: LCCN 2017040534 (print) | LCCN 2017040885 (ebook) | ISBN
  9781604277920 (e-book) | ISBN 9781604271492 (hardcover : alk. paper)
Subjects: LCSH: Project management. | Information technology—Management.
Classification: LCC HD69.P75 (ebook) | LCC HD69.P75 K85 2017 (print) | DDC
  658.4/04—dc23
LC record available at https://lccn.loc.gov/2017040534

Phone: (954) 727-9333
Fax: (561) 892-0700
Web: www.jrosspub.com

CONTENTS

FOREWORD

It gives me great pleasure to write the foreword for Murali Kulathumani's new book, *Breakthrough Project Portfolio Management*. Throughout my career in information technology and presently as chief information officer (CIO) of a Fortune 500 company, I have always found the discipline of portfolio management to be a valuable aid in choosing and monitoring investment choices. I've found that portfolio management can make the difference between mere project activity and meaningful achievement of capability.

Part of the challenge for me has been to visualize a road map of capabilities for the discipline of portfolio management so that we can plan to take our organizations to the next level of maturity. In my opinion, while there are many good volumes about portfolio management in the market today, this new book is one of the first to propose an easy-to-use model of portfolio capabilities and the different levels of attainment.

Another aspect of portfolio management where I have felt there could be a more detailed exploration has been the gap between portfolio theory and the realities or constraints encountered in real life. While the theory of portfolio management is well established, there remains a considerable gap in applying this theory to the varied situations that are encountered in organizations today. Every organization is in a different place in terms of process maturity, and I believe there is always a need to address the gap between theory and practice. This is where this book plays a valuable role in making the reader aware of how to navigate the space between the *ideal* and the *real*, while delivering on the promise of portfolio management.

A third aspect of this readable book that appealed to me was its adaptation of earned value management tailored for portfolio management. While the discipline of earned value management has been a proven workhorse in the defense and construction industries, the perception of high rigor and effort around this technique has prevented it from being more widely adopted. This book has approached this standard technique in a new way, distilling the most essential aspects into a simple set of artifacts that should be easy for most organizations to use. The use of this modified form of earned value management to serve as an objective indicator for project and portfolio performance is useful in providing executive decision makers with the tools to make data-driven choices. The book not only introduces this simplified technique, but also goes on to build upon that same foundation, an extensive methodology to aggregate data for easy visualization by executives. I also found it valuable and thought provoking to peruse the detailed chapter on providing concrete metrics on transformational strategy attainment using the modified earned value method.

From a CIO's perspective, perhaps the most compelling feature of this book is the detailed exploration of how the portfolio office and the CIO are connected in the delivery of value to the organization. Several aspects of this topic gave me pause for thought and are worth exploring for any CIO to optimize the output from their portfolio office.

Although dealing with technical topics and detailed methodologies by which to approach these topics, the book has been written in an easy-to-follow manner, with handy illustrations aiding readers in their comprehension of the ideas and techniques. I believe this book stands apart from similar works in the field and that it is a must read for anyone interested in portfolio management, as well as decision makers who look to achieve strategic transformation through portfolio management.

Sean Perry
CIO, Robert Half International

*This book is dedicated to my dear family,
Appavu, Abirami, Vidya, and to my dear parents."*

PREFACE

"If there's a book you really want to read but it hasn't been written yet, then you must write it."

—Toni Morrison

The genesis of this book could be characterized along similar lines to the above quote. As a practicing portfolio manager, I was always in search of books that could help me improve my craft. Barring a few exceptions, I was always stymied by the chasm between theory and practice; by the difference between the idealized treatment of topics in the portfolio management books and the cold hard reality of the workplace.

Eventually, I devised mechanisms that helped me bridge this divide and make portfolio management successful in the different companies where I had the opportunity to work. However, I realized that this knowledge deserved a wider reach and that this could be an opportunity for a new generation of portfolio managers who were grappling with the very same issues that I had solved through trial and error. That realization eventually led to the creation of the book you hold in your hands now.

I had three main goals in writing this text:

- Sequentially cover the essential capabilities of an effective portfolio and provide a logical construct of how the different capability areas interact
- Provide a complete understanding of all the building blocks of a portfolio, the critical success factors needed to achieve desired results, and the nuances involved in implementing them
- Introduce a pioneering *mEVM* methodology and articulate various strategies for implementing it within an organization

Having covered the above in sufficient detail, I felt that there was still something to be said about the "team sport" nature of portfolio management. An effective portfolio always works in partnership with other functions such as Finance, the office of the CIO, and the business. And it usually takes more than one person to implement an effective portfolio—it takes a whole team (hence the chapter on the portfolio office).

Most books, and certainly this one, do not take shape in a vacuum. A whole host of people were instrumental in the creation of this text. Foremost among them are mentioned below, but many more were involved in the interactions that enabled this text. First, Drew Gierman at J. Ross Publishing, for taking a chance on a first-time author and his ample patience in waiting for me to deliver the promised manuscript. Thanks are due to Sean Perry, CIO of Robert Half, who is the kind of CIO a portfolio manager would love to partner with and work for. Much of the chapter on the CIO's role is modeled on my interactions with Sean. The community of portfolio managers at Robert Half also deserve thanks for their collegial spirit in helping each other and displaying a high degree of competence in this field.

When it comes to the art of portfolio management, mention must be made of Jennifer Cheng, Sr. Director at Kaiser Permanente, who exemplifies many aspects of the ideal portfolio manager. I learned many useful portfolio management techniques from Jennifer and her encouragement was foundational in my creation of the mEVM technique. I would also like to thank Fidelis Atuegbu, former Director at Kaiser Permanente, for showcasing the ideal partnership between the finance department and the portfolio office.

Admiration and thanks go out to Murali Chemuturi and Dr. Prasad Kodukula, both stalwart fellow authors in the J. Ross Publishing family, for the inspiration I received from them. Finally, thanks are due to the production staff at J. Ross Publishing, specifically Jackie Lininger and Steve Buda. Their patient and meticulous edits turned my sometimes telegraphic prose into meaningful and readable content.

Ultimately, portfolio management is the "art of the possible". In the face of change, in the face of adverse developments and unforeseen risks, the portfolio manager is still expected to navigate the portfolio to the safe harbor of impactful strategic results. If this book assists in that endeavor, I will consider my efforts to be successful.

Murali Kulathumani

HOW TO USE THIS BOOK

INTRODUCTION

The aim of this book is to enable you to transform your current portfolio into a world-class portfolio. Whether you already have a portfolio, or are starting to build a portfolio from scratch, the contents of this book will inform you about the capabilities of a high-performing portfolio and help you get there. The goal is to provide the reader with a complete understanding of all the building blocks of the portfolio and then understand the nuances involved in implementing the same. The chapters, which sequentially cover the essential capabilities of a portfolio, are structured in a simple, intuitive way and also cross-reference each other to provide the reader with a logical construct of how the different capability areas interact.

PREREQUISITES

This book assumes very little in terms of prerequisites on the part of the reader. A basic knowledge of projects, coupled with a passing understanding of finance terms and modern organizations are all that it takes for a reader to understand and start implementing the concepts explored in this book.

OVERVIEW OF THE BOOK STRUCTURE

This book is divided into four parts. Part I covers all of the key components of a portfolio management process. Chapter 1 describes the mission-critical role of the high-performing portfolio. Chapters 2 and 3 deal with the mechanics of managing the portfolio intake and orchestration of the annual planning process, respectively. Chapter 4 delves into funding strategies for the portfolio, while Chapter 5 deals with monitoring the performance of the portfolio and its constituents. Chapter 6 covers the important task of balancing the portfolio and Chapter 7 concludes Part I by exploring the management of realization of project benefits.

Part II of this book covers a central theme of the book—namely, the utilization of a simplified version of earned value management (EVM) to objectively measure and manage a portfolio of projects. Chapter 8 starts off with an introduction to the concept of modified EVM (*mEVM*) and Chapter 9 builds on the concept with more tactile artifacts of the mEVM system. Chapter 10 introduces the concept of data aggregation and describes a variety of scenarios where this could be deployed for the benefit of portfolio governance and other executives. Chapter 11 takes an unconventional approach to measuring strategic attainment using the mEVM concept, while Chapter 12 addresses the implementation of mEVM by describing various strategies to roll out mEVM to the enterprise successfully. Finally, Chapter 13 rounds out Part II by explaining why mEVM works.

There is a world of difference between theory and real life. Part III of this book grapples with implementation strategies for the real world and starts with Chapter 14, which directly addresses the most common problems faced by portfolio managers as they try to roll out capability enhancements in their organizations. Chapter 15 underscores the importance of having systems and tools that actually work in making portfolio management possible. Chapter 16 addresses the dominant factor that can make or mar portfolio performance—namely, the politics at work in organizations—and how to successfully navigate those politics. Chapter 17 brings Part III to a close with a detailed look at portfolio governance and how to ensure that it successfully provides direction for the portfolio.

No successful portfolio operates in a vacuum. Part IV explores in depth the support systems that play a huge role in making the portfolio successful. Chapter 18 explores the important relationship between the chief information officer and the portfolio office, while Chapter 19 follows up on the relationship between the finance department and portfolio office. Chapter 20 highlights the critical role played by the change management function in preparing the organization for changes rolled out by the portfolio office. Chapter 21 explores the characteristics of an ideal portfolio manager, strategies for staffing the portfolio office, the typical composition of a portfolio office, and the ideal reporting structure and place in the organizational hierarchy. Chapter 23 concludes Part IV with an exploration of the role played by the business in enabling the success of portfolio management.

CHAPTER STRUCTURE

Every chapter begins with an introduction to the central topic of that chapter. As an element of the introduction, a summary listing of the chapter's contents is provided to enable the reader to get a bearing as to how the chapter unfolds. This is typically followed by another section that elaborates on the introduction with an informative, more detailed discussion. Some chapters will also have a section that describes the need for the topic that chapter is based upon. Several chapters employ the technique of progressive elaboration of the topic at hand, using tables and diagrams as appropriate. For most chapters, there typically follows an explanatory section that deals with how to set up the building blocks of a certain capability. Finally, there is also a section that describes the levels of capability maturity for that topic and the attendant characteristics of each level. The chapter summary provides a synopsis of each chapter, and at the end of the chapter is a listing of references or notes, if any, to sourced content used in that chapter.

COMPARISON WITH FINANCIAL PORTFOLIO MANAGEMENT

A singular difference between this book and most other volumes on portfolio management is the use of financial portfolio management to introduce some topics in project portfolio management. The author believes that most people are familiar with financial portfolios for the simple reason that they are likely to own one or more. It is therefore reasonable to expect people to grasp project portfolio concepts when they are introduced as a variant of the already familiar financial portfolio concepts. However, this comparison is applied judiciously and, where appropriate, the differences between the financial and project portfolio concepts are highlighted.

THE PORTFOLIO OFFICE AND THE PORTFOLIO MANAGER

Although the portfolio office consists of more than just the portfolio manager, it needs to be remembered that many organizations only have one person—namely, the portfolio manager, managing the portfolio. This is especially true for organizations that are still starting on their portfolio journey. Where there is a larger portfolio office, the portfolio manager is understood to be the prime driver within the portfolio office and that the other members of the portfolio office function under his or her direction. Therefore, the two entities of portfolio office and portfolio manager are used interchangeably, unless expressly indicated otherwise.

FOCUS ON mEVM

Although this book can be used as a complete text on portfolio management, Part II—which covers mEVM—can be used independently for portfolios that already have a sufficient measure of capability in place. With its detailed chapters on the subject—from introductory to advanced uses—Part II contains all of the necessary information for an organization to adopt this powerful technique and to transform their portfolio to the next level.

THE CONTINUOUS JOURNEY OF PORTFOLIO MANAGEMENT

Every portfolio is at a different level in terms of capability as a result of many factors, including the context of the larger organization. Consequently, it's natural that every portfolio manager will approach this book a little differently, based on their current place in the journey. To aid in this approach, references have been inserted in each chapter that enables the reader to look up other chapters where a topic may have been explored in greater depth.

ACCESS TO TEMPLATES

This book was written with an emphasis on impactful implementation in the real world. Accordingly, the tables and figures used in the various chapters have been made available in their original form as a resource to jump-start the reader's implementation journey.

CONCLUSION

Portfolio management can be a challenging endeavor. It can also be a rewarding journey—especially as the organization begins to become aware of the potential of this field of application. This book tries to enable the readers and their organizations to become successful in that journey by listing out the different components that make a portfolio work as well as the subtle nuances that have been proven effective by observation and experience. It is the fervent hope of the author that the readers are able to deploy the content of this book to create high-performing portfolios in their own organizations.

Murali Kulathumani

ABOUT THE AUTHOR

Murali Kulathumani, PMP, has over 20 years of information technology management experience. He has successfully managed large project portfolios at leading Silicon Valley firms such as Cisco, Symantec, and Kaiser Permanente. Murali has extensive experience with the full spectrum of portfolio capabilities. This leading expert also pioneered a simplified form of earned value management, called *mEVM*, which has been well received by industry practitioners and various organizations. In fact, it has become the standard at a billion-dollar business unit of a leading health care provider in the United States.

Mr. Kulathumani has a technical degree in Electrical Engineering from Bangalore University and an MBA from Purdue University. Murali earned the Project Management Professional (PMP)® designation from the Project Management Institute and he is a Certified Scrum Master. He is a published author, consultant, trainer, professional speaker, and has also taught courses as an adjunct professor at Purdue University and the University of Phoenix.

Murali has witnessed firsthand the turnarounds that can take place with proper guidance. This motivated him to share his expertise and to formalize the body of knowledge that he has gained in the form of a book that explains in detail how to effectively manage project portfolios in real-world practice. His goal is to help the many practitioners and companies that have been struggling to implement portfolio management simply because there hasn't been enough guidance about which success factors deliver the greatest impact.

At J. Ross Publishing we are committed to providing today's professional with practical, hands-on tools that enhance the learning experience and give readers an opportunity to apply what they have learned. That is why we offer free ancillary materials available for download on this book and all participating Web Added Value™ publications. These online resources may include interactive versions of material that appears in the book or supplemental templates, worksheets, models, plans, case studies, proposals, spreadsheets and assessment tools, among other things. Whenever you see the WAV™ symbol in any of our publications, it means bonus materials accompany the book and are available from the Web Added Value Download Resource Center at www.jrosspub.com.

Downloads for *Breakthrough Project Portfolio Management* include various tables and figures found in the text and a set of Excel files which constitute the Capability Maturity Tool Kit.

Part I

Key Components of a Portfolio Process

1

THE MISSION-CRITICAL ROLE OF THE HIGH-PERFORMING PORTFOLIO

INTRODUCTION

A chief information officer's mandate is essentially that of a change agent. He or she is expected to effect a change for the better in systems, performance, and capabilities of the organization. The time-tested way to effect change is through a well-managed portfolio of projects—a project being a temporary endeavor that results in an end product, capability, or service.

However, a well-managed project portfolio is one of those mythical information technology (IT) beasts—often bragged about and rarely bagged. An optimized, high-performing portfolio is an ensemble of optimized parts, coming together to deliver a whole that is greater than the sum of its parts. This chapter will explore the characteristics of an optimized portfolio and outline the strategic impact of having such a portfolio. It will also depict the key components of an optimized portfolio, which are then elaborated on throughout the rest of Part I. This chapter also delves into the importance of the maturity level in each portfolio capability along with the net impact of portfolio capability maturity on the pace of strategic transformation.

CHARACTERISTICS OF THE OPTIMIZED PORTFOLIO

An optimized, high-performing portfolio has the following characteristics:

- *Transformative*—The portfolio is able to describe and execute a road map from where the organization is today to where it needs to be tomorrow. It's able to manage all the components and show what they add to the strategic journey from the capabilities of today to those of tomorrow.

- *Adaptive*—The portfolio is able to deal with changes in funding and direction. It's able to pick up additional funds as they become available and put them to work in furthering the strategic agenda. In a dynamic business environment, it enables the organization to pivot as necessary.
- *Transparent*—The portfolio can show all the projects and what their key parameters are at a glance. It enables decision makers to make the right decisions based on data.
- *Efficient*—The portfolio allocates resources to projects in the most efficient way possible. It maximizes the benefits accrued to the organization by ensuring optimal distribution and use of resources.
- *Predictive*—Beyond the factual reporting of current and historical project performance, a high-performing portfolio is able to look ahead and predict what is likely to happen, giving executive management a distinct advantage in making the right decisions.
- *User Friendly*—A high-performing portfolio is also a user-friendly system. It makes every user interaction with the system as quick and painless as possible.
- *Impactful*—A well-run portfolio can clearly demonstrate the value delivered by the portfolio both to IT and the business. It is also able to show progress along the strategic road map and quantify strategic attainment and correlate that to the resources expended.
- *Operationally Sound*—For a portfolio to do all of the above, it must be run well on a day-to-day basis. The portfolio office should be equipped with the right tools, staffed well, and be responsive to the needs of the stakeholders and the organization.

THE STRATEGIC EFFECT OF A WELL-RUN PORTFOLIO

Let's look at what a portfolio adds to an organization at a high level—beyond the management of dollars and cents. Figure 1.1 is a simple graph with the y-axis depicting the strategic capability/scale/capacity of an organization; while the x-axis depicts a combination of time and money. The graph also shows the organization's modest strategic capability at the present (marked as *Today*). Figure 1.1 also depicts a high-performing portfolio at work, managing a stream of well-chosen projects that align with the strategic direction of the organization. Furthermore, the portfolio can spot and remediate projects in trouble. While every project is managed by a project manager, the portfolio office is able to harness the transformative power of a portfolio and ensure that, over time, the organization will make significant strides in its strategic road map to achieving new capability, scale, and capacity. This *new* state of the enterprise's strategic capability in the future (marked as *Tomorrow* on the x-axis) is shown as the bigger tower representing enhanced capability and capacity.

In contrast, Figure 1.2 depicts the same amount of time and resources spent in pursuing a disparate collection of projects that are not managed under an effective portfolio. Although every project is still managed by a project manager, there isn't the overarching oversight provided by a portfolio office, thus, projects are left to police themselves. In this case, it's hard to predict the net effect of the projects in achieving strategic goals, but in all likelihood, the net output over

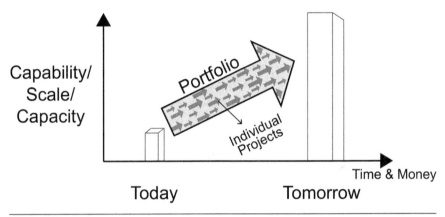

Figure 1.1   Growth in strategic capability over time when projects are managed in a high-performing portfolio

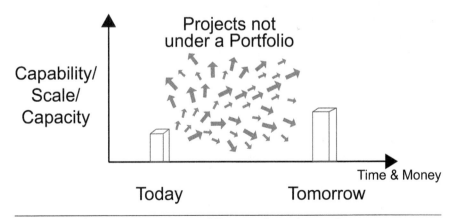

Figure 1.2   Growth in strategic capability over time when projects are *not* managed in a high-performing portfolio

time will be far more modest, as shown at the far right of Figure 1.2. Without a portfolio office, strategic goals could still be achieved, but definitely as a more random function—what's undeniably missing is the predictability, control, and assurance that is obtained by having an effective portfolio in place. The biggest risk in Figure 1.2, from an executive's point of view, is that there is no feedback about the pace of the strategic journey and, hence, little room for correction until it's too late.

THE CONCEPT OF PORTFOLIO CAPABILITY MATURITY

Given the audience of this book, the importance of portfolio management is an accepted conclusion. The widely shared view is that portfolio management is indispensable in today's organization—whether business or IT. Accordingly, most organizations have a portfolio or are in the process of implementing one. But, of course, all portfolios are not equally successful and this leads to the question of what it takes to assemble an effective portfolio. Here we introduce the concept of portfolio capability and the different levels of maturity in each capability. The following section explores the concept of portfolio capability and also illustrates how the level of maturity in each capability can make a difference in the overall effectiveness of the portfolio from a strategic transformation perspective.

Figure 1.3 shows a graph with each axis representing a particular portfolio capability. For each capability, there are levels that denote the respective maturity of that capability. For example, Level 1 is indicative of a modest or rudimentary achievement in that capability, while Level 3 represents a high degree of attainment in that capability. It's also worth noting that a portfolio may have different levels of attainment in different capabilities.

Against this backdrop, let's consider two portfolios—Portfolio A and Portfolio B. Portfolio A has a Level 2 of attainment in Capability 1, but only a Level 1 in Capabilities 2, 3, and 4. Connecting these points gives us a quadrilateral that is shown in Figure 1.4 by a dotted outline. This dotted-line quadrilateral is then mapped over to a *pipe* of IT strategic transformation, with the outline of the quadrilateral forming an aperture of the pipe. Figure 1.5 shows how the quadrilateral formed by the levels of Portfolio A maps into the aperture of the IT strategic transformation pipe. We see that the modest capability of this portfolio creates a modest aperture in the pipe of IT strategic transformation.

What would happen if a portfolio had high levels of attainment in every capability? Consider Portfolio B, another portfolio which has an attainment of

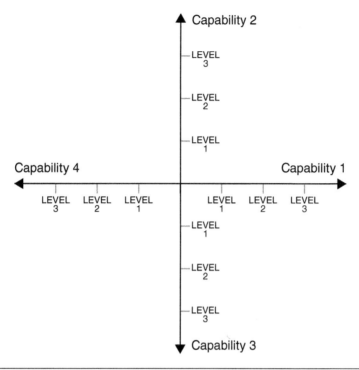

Figure 1.3 Graph with each axis representing a particular portfolio capability

Level 3 in Capabilities 1, 2, 3, and 4 and could be termed as a high-performing portfolio in all respects. As in the previous example, a quadrilateral is formed by connecting these points and then mapped over to the pipe of IT strategic transformation. First, we see that a much bigger quadrilateral is formed in the case of Portfolio B when the portfolio has a Level 3 attainment in each capability. Second, we see that this bigger quadrilateral maps into a correspondingly bigger aperture in the transformation pipe. Figure 1.6 shows the mapping of the quadrilateral formed by the capability levels of Portfolio B into the aperture of the IT strategic transformation pipe.

What is the impact of having a bigger aperture in the strategic pipe? Simply put, a bigger aperture translates to a larger amount of strategic throughput. Figure 1.7 examines the strategic throughput produced by the two portfolios—A and B. We can see that Portfolio B, the high-performing portfolio, has a significantly higher strategic throughput. The takeaway is that higher levels of attainment in each portfolio capability correlate directly with increased strategic throughput and value delivered to the organization.

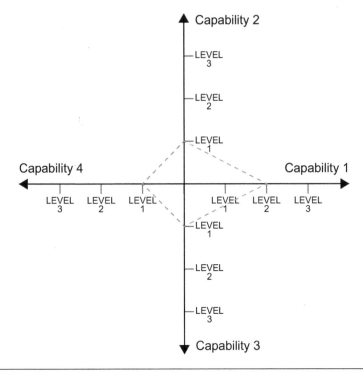

Figure 1.4   The connection of capability levels of Portfolio A to form a quadrilateral

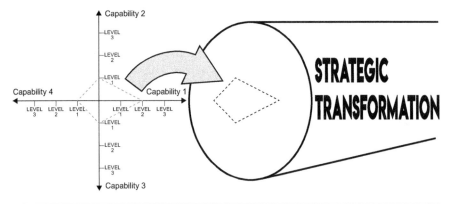

Figure 1.5   The quadrilateral formed by the levels of Portfolio A into the aperture of the IT strategic transformation pipe

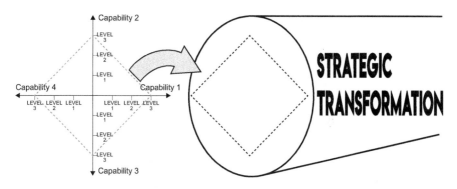

Figure 1.6   Mapping of the quadrilateral formed by the levels of Portfolio B into the aperture of the IT strategic transformation pipe

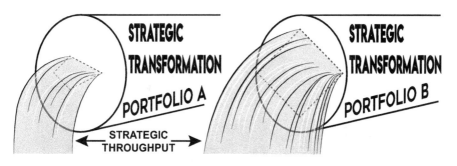

Figure 1.7   Comparing the strategic throughput produced by the two portfolios—A and B

RELATIONSHIP OF THE MATURITY OF PORTFOLIO COMPONENTS TO OVERALL PORTFOLIO MATURITY

Figure 1.8 shows how the capability of the portfolio components affects the overall capability of the portfolio. The important components of the portfolio are as follows:

- Portfolio intake management
- Portfolio funding management
- Portfolio benefits management
- Modified earned value management (*mEVM*) process management
- Portfolio rebalancing management

- Portfolio information systems
- Portfolio annual planning management
- Portfolio rollout/change management

Each of these is explored in detail in respective chapters in this book. Also, note that each of the component processes have levels of maturity from Level 1 to Level 3. From Figure 1.8, it can be seen that the level of maturity of the overall portfolio (the big cylinder in the center) is a function of the levels of maturity of the smaller cylinders in the periphery. When the portfolio processes represented by the peripheral cylinders tend to have a higher level of capability, so will the overall portfolio represented by the central cylinder. The key to a high-performing portfolio is to increase the levels of maturity in each of the portfolio processes.

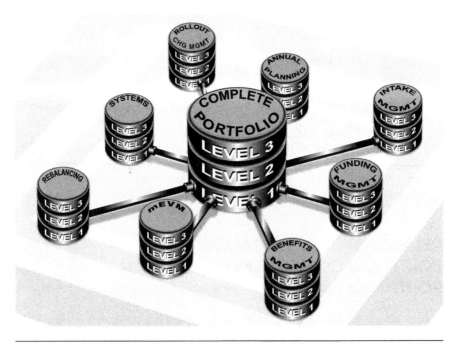

Figure 1.8   How the capability of the portfolio components affects the overall capability of the portfolio

This book has free material available for download from the
Web Added Value™ resource center at *www.jrosspub.com*

2

PORTFOLIO INTAKE AND ASSESSMENT

INTRODUCTION

Portfolio intake is the process of regulating the work that enters the portfolio. It's vital to get this working correctly because the portfolio should be working only on proposals aligned with the strategic direction of the portfolio and not waste precious resources on misaligned proposals. However, the goal of getting the portfolio to work on the right things also needs to be balanced with the goal of maintaining speedy throughput—namely, getting the right things done quickly. In this chapter, we will cover the following aspects of portfolio intake management:

1. Define how an ideal portfolio intake should work
2. Specify the two broad types of portfolio and detail how the portfolio intake will vary depending on the portfolio
3. Explore the challenges in administering the intake in each kind of portfolio
4. Explore the role of annual planning as a primary feeder for the intake process
5. Finally, we'll round out with some best practices that work regardless what the organizations have in place; the goal is to have robust intake management while preserving the portfolio's responsiveness and speed

THE IDEAL PORTFOLIO INTAKE

How should an ideal project portfolio intake function? To arrive at that vision, let's start with a useful analog: the intake for a financial portfolio. (By the way, this is a recurrent theme in several chapters of this book—we will use the financial portfolio as an introduction to the various functions of a project portfolio.) So how does the intake function in a well-run financial portfolio?

A financial portfolio portrays a very clear vision of what its strategy is and, by extension, what kinds of investments should be included in the portfolio. For example, the portfolio may be focused on large-cap companies. In that case, there would be a *filter* that selects only large-cap companies for possible inclusion in the portfolio. In addition, there would be additional entry criteria for potential investments such as price/earnings ratio, debt ratios, management track record, etc.

When similar filters are applied to the project portfolio, the previous guidelines would translate to the following considerations:

- **Strategic fit**: How is this proposal a fit for the goals of this portfolio? (Only applies if this is a strategy-focused portfolio.)
- **Cost**: What does this cost in terms of operating expense and capital expense—by quarter, by year?
- **Strategic road map impact**: How does this project relate to our strategy? What capabilities will this project add to our strategic road map? (Only applies if this is a strategy-focused portfolio.)
- **Benefits**: What is the return on investment? Are the benefits soft (productivity savings, etc.) or hard (revenue, actual cost reduction in the general ledger, etc.)?
- **Resources needed**: What are the estimated resource requirements for this project?
- **Opportunity cost**: What is the cost of not doing this project? (This is good to have in case the proposal is not approved for funding.)

Only the proposals that meet the listed criteria are considered for funding and inclusion in the project portfolio. An ideal portfolio intake has a filter in place that applies these criteria and selects the right proposals. In addition to a well-defined filter, the ideal portfolio intake also has the following characteristics:

1. **Efficient and responsive**: Approval or non-approval should be expedited. A well-defined cadence should be in place to make intake decisions.
2. **Easy to use**: The organization should find it easy to submit a complete proposal. There should be resources and training that enable an organization to know what to do when submitting a proposal. At the same time, only complete proposals should be submitted to the intake.
3. **Effective governance**: All intake decisions should be overseen by the appropriate governance body that has been empowered to approve/deny projects in accordance with the goals of the portfolio.
4. **Fiscally responsible**: Prior to approving a project, the governance body should have a clear idea of what this does to the available portfolio funds.

5. **Proportionally rigorous**: Portfolio intake should have variable paths and speeds of processing, depending on the types of proposals. Previously known, *vetted* proposals should be expedited through the process, while appropriate rigor should still be applied to new proposals.

PORTFOLIO TYPE AND ITS IMPACT ON PORTFOLIO INTAKE

Although we outlined the characteristics of an ideal portfolio process in the previous text, in reality, the intake process will be heavily influenced by the kind of portfolio. Three broad portfolio types are defined below:

1. The transformational or strategic portfolio
2. The tactical or functional portfolio
3. The hybrid portfolio

The Transformational or Strategic Portfolio

A transformational or strategic portfolio is one that seeks to transform the capabilities of the organization in line with the strategic vision. It may seek to significantly change or enhance the positioning of the organization in the marketplace vis-à-vis its competitors and/or enable the organization to enter new markets. The innovation portfolio to be discussed later is a subset of the transformational portfolio.

A well-defined strategy is key to a transformational portfolio. "Transform to what?" is the key question that needs to be addressed. It's not within the scope of this book to go into how to formulate strategy—we'll instead focus on where the formulated strategy interacts with the portfolio intake. This is also covered in more detail in Chapter 3.

Strategy formulation is a popular endeavor in organizations. Few organizations would confess to not doing strategy at all. However, all too often, strategy formulation efforts tend to result in the production of high-level aspirational strategy (aka *Credenzaware*)—but for portfolio intake purposes, it's important to decompose the high-level strategy into an objective list of factors that can be aligned with the portfolio.

Table 2.1 depicts an artifact of portfolio management called the strategy decomposition table. It is an important artifact which interfaces with many facets of portfolio management and will be used in several chapters that deal with different areas of portfolio management.

Here is a brief introduction to the table: Column 1 of the table contains the high-level organizational strategy, which is then decomposed into a set of

Table 2.1   A basic form of the strategic decomposition table

High-Level Strategy	IT Strategic Priority	Description of Priority	Sub-projects Aligned to the Priority
Aspirational: Vision for the Future	1	Description of Priority #1	Sub-project 1
			Sub-project 2
			Sub-project 3
	2	Description of Priority #2	Sub-project 4
			Sub-project 5
			Sub-project 6
	3	Description of Priority #3	Sub-project 7
			Sub-project 8
			Sub-project 9
	4	Description of Priority #4	Sub-project 10
			Sub-project 11
			Sub-project 12

strategic priorities (Column 2), which are then subsequently decomposed into a set of sub-projects (Column 3) with concrete scope, schedule, and budget (not shown here). Although the strategy table shown in Table 2.1 is basic, it will be expanded upon with additional columns and details in later chapters.

Operation of a Strategic Portfolio Intake

Given the framework as described in the strategy table, here's how the strategic portfolio's intake would work, in a two-tiered arrangement:

Tier 1: Qualification of the Proposal

- When a proposal is received, the portfolio manager matches it with one or more strategic priorities. The proposal owner could also specify what strategic priorities their proposal aligns with.
- The link to the priority must be validated by the priority advocate for that strategic priority. (The role of a priority advocate is explained in Chapter 3.) This validation by the priority advocate is necessary because people are often not clear about what priority their project aligns to. This validation also guards against a scenario where some stakeholders try to project a false alignment to important strategies hoping to secure approval and funding for their project.
- There needs to be a clear specification of what this project is expected to contribute in terms of the strategic road map. This clear specification

should preferably be accompanied by the strategic road map (refer to Chapter 3 for more details).

- Calculation is done as to whether the portfolio can afford to do this project and where the financial reserves would stand if this project was approved. If there are a batch of projects being considered in this round of intake, the projection should show what the combined cumulative financial impact would be (of approving all the projects). This can spur a productive discussion about approving some projects and rejecting others, at least for the time being.

Tier 2: Review by Governance

What is portfolio governance? Portfolio governance is a team of senior people providing oversight and decision making for a portfolio. They are the body that approves (or denies approval of) key decisions regarding projects in the portfolio. Portfolio governance extends to intake since it's important to screen new proposals seeking to be a part of the portfolio. Governance performs the following actions on intake:

- Reviews fully developed proposals including estimates of effort and expense
- Determines financial impact of approving one or more proposals
- Makes final go/no-go decision on each proposal
- Sends back some proposals with requests for more information that are then reconsidered at subsequent sittings

Intake Considerations of a Strategic Portfolio

The following considerations are the most important for a strategic portfolio:

- **Modest volume**: Since the efforts are fairly large and transformational, the volume is likely to be modest compared to a functional portfolio
- **Need for a high degree of confidence in strategic alignment**: In a strategic portfolio, the emphasis is on only doing projects that are in alignment with the strategic mission of the portfolio. Therefore, before a proposal is cleared for funding and inclusion in the portfolio, there needs to be a high degree of confidence in the alignment of the proposal to the strategic road map.
- **Expectation to spend effort in validating alignment**: In order to perform all the necessary validation to ensure alignment, it takes a fair amount of time and effort. A strategic portfolio's intake is not necessarily fast moving.

The Functional Portfolio

What is a functional portfolio? A functional portfolio is one that aims to orchestrate the efficient functioning of a department or organization by managing a mix of mostly operational projects. It is not necessarily trying to transform the capabilities of the organization—it operates at a more tactical level. In other words, a functional portfolio has a more utilitarian goal in mind; it operates in a space where the big picture has already been defined and the strategic direction has already been established. The emphasis here is on maximizing the organization's throughput while ensuring that money is spent wisely.

Examples of projects that are found in a functional portfolio include:

1. Upgrading the version of an operating system installed on the companies' laptops
2. Replacing end-of-life systems and software
3. Making configuration changes to an enterprise resource planning (ERP) module
4. Updating the company's website, etc.

All the previously listed efforts are important, yet unlikely *by themselves* to bring about dramatic change in the organization's capabilities or result in strategic transformation.

Operation of a Functional Portfolio's Intake

The portfolio team does the following actions in managing the intake of a functional portfolio:

- All proposals are submitted using a common tool and/or standard format (see Chapter 15 for more details on tool recommendations).
- The portfolio manager holds a periodic meeting (weekly or biweekly is recommended) to review proposals.
- A representative group of leaders or their representatives are invited for the meeting.
- One vital output of this meeting is the identification of redundant and/or duplicate proposals (example: writing a set of reports for an ERP module that is being retired/made obsolete).
- Another output would be the identification of cross-team impacts. Questions that need to be answered include: Who would be impacted by this project? Which team(s) need to contribute resources for this project?
- All the teams that are impacted are expected to state their level of effort within a service-level agreement (SLA)—say two weeks.
- The combined cost of these efforts are rolled into the final estimate for the proposal; and funding approval is secured from the executive owning the functional portfolio.

- The functional portfolio manager needs to know who the main stakeholders are—for example, he or she would know that a marketing campaign proposal from the marketing department would not belong in her IT portfolio and needs to be screened out.

It should be noted that set up in a functional portfolio is simpler—this is because there is no strategy aspect to validate and align to.

Intake Considerations of a Functional Portfolio

The following considerations are most important for a functional portfolio:

- **High volume**: In a functional portfolio, the volume is likely to be huge since each effort is quite small.
- **Need for speedy throughput**: In order for the functional portfolio to be effective, it needs to operate in a way that doesn't slow things down.
- **Alignment is not a major concern**: Unlike a strategic portfolio, misalignment with strategic priorities is not really applicable here. The screening process is more of answering yes or no to the following question: Does this effort belong in our portfolio or not?

The Hybrid Portfolio

In large organizations, there are sufficient volume and funds to justify specialized portfolios such as a purely strategic or purely functional portfolio. In smaller portfolios, or in portfolios that are just being started, it's more likely that the portfolio manager is expected to manage both functional and strategic proposals. How should a portfolio manager approach a hybrid portfolio? Here are the key pointers to success in a hybrid setup:

1. A hybrid portfolio needs clear lanes of separation between functional work and strategic work—the key reason for that being the differential speed of processing for each type of work.
2. Functional work is small, rapid, and voluminous—it should not get held up due to the slower strategic work.
3. Strategic work is costlier, slower to implement, and not as numerous—it needs a full validation and alignment check to ensure that the portfolio funds are committed wisely.

Operation of the Hybrid Portfolio's Intake

- All proposals are submitted using a common tool; however, there are instructions provided to enable the submitter to self-classify their proposal as functional (simpler effort) or strategic (complex effort).

- Depending on the choice made (functional versus strategic), the user has to fill out a different set of fields. The functional proposals have fewer fields and the strategic proposals have a more extensive set of fields.
- The fields describing the strategic work are identical (or very close) to the fields used in the annual planning template (see Chapter 3 for more details regarding the annual planning template).
- The portfolio manager reviews proposals to make sure that the users identified the work correctly (functional versus strategic) and filled out the correct information.
- The portfolio manager holds a periodic meeting (weekly or biweekly is recommended) to review the proposals.
- Based on whether the proposal is functional or strategic, the portfolio manager communicates how the proposal will move forward.
- If the proposal is functional, the steps will be the same as outlined in the functional portfolio:
  1. This proposal is added to the agenda of the standing meeting to which a representative group of leaders or their representatives are invited.
  2. Determination is made if the proposal is redundant or a duplicate.
  3. Cross-team impacts are identified. Who would be impacted by this project? Which team(s) need to contribute resources for this project?
  4. All the teams impacted are expected to state their level of effort within an SLA—say two weeks.
  5. The combined cost of these efforts is rolled into the final estimate for the proposal, and funding approval is secured from the executive owning the functional portfolio.
- If the proposal is strategic, the steps will be the same as outlined in the strategic portfolio:
  1. The proposal is validated as a bona fide strategic proposal that is in alignment with the strategic road map. In order to perform this validation, the portfolio manager may need to consult the respective priority advocate (the role of the priority advocate is explained in Chapter 3).
- Once the proposal is validated, the portfolio governance body reviews the proposal including estimates of effort and expense. It determines the financial impact of approving one or more proposals and makes the final decision on each proposal.

ROLE OF ANNUAL PLANNING

While considering how to best manage portfolio intake, the source of the project proposals should also be considered. Proposals are mainly identified during an annual planning exercise simply called *annual planning* (or a close sounding variant). It is an annual effort to put all known demand on the table from every department in the organization. The detailed mechanics of how to orchestrate and manage the annual planning cycle is covered in Chapter 2, but for the purposes of this chapter, we explore how the annual planning process works with the intake process.

- Annual planning involves analysis about all the proposals brought forth—this analysis is done using the planning artifacts that accompany each project. At the end of this analysis, a list of prioritized projects are created and the topmost projects in that list are funded and approved for execution in the next year.
- The projects that are further down on that list, but not funded (for lack of money), are added to a *queue*. In other words, these projects are still viable investments, but are currently waiting for funds to become available.

Now a key question to be considered is whether the proposals already seen at annual planning and approved for execution still need to come through the intake. The recommended best practice is that all proposals come through the intake process, for trackability and governance purposes. However, proposals previously approved through annual planning are able to use an *expedited track* as described here:

- The approved projects are allowed to reuse their annual planning artifacts to expedite processing and avoid duplication of data. This also underscores why the intake process needs to mirror the information collected during annual planning.
- Strategic validation and governance approval also happen, but are done more as a formality without additional work on the part of the proposal owners. (This is because the strategic validation has already occurred during annual planning. It may become necessary for the portfolio team to confirm that this has indeed occurred for the proposal in question.)

What about the rest of the projects that are on the *queue*? The queue of projects becomes a potential source of proposals when funding is freed up and new proposals are sought. When these proposals are ready to come forward to seek funding, they can reuse the annual planning artifacts. Of course, these proposals need to come through the intake process as well.

In summary, annual planning provides a steady stream of vetted proposals that are fed into portfolio intake. Annual planning is closely related to the strategic planning and the associated strategic road map, all of which are important components used in the intake process. A high-performing portfolio usually has a robust annual planning function as a complement.

BUILDING AN INTAKE SOLUTION USING A WORD/EXCEL TEMPLATE

There are a couple of methods for implementing a basic intake solution. The simplest method is to create a standard intake form in Word or Excel and distribute it by e-mail.

- **Key fields to include in the Word/Excel document (for both functional and strategic proposals):**
  - Proposal number
  - Description
  - Whether seen/approved at annual planning
  - Rough estimate of cost and duration
  - Risk of doing
  - Risk of not doing

- **Additional fields to include for strategic proposals only:**
  - Specify which strategic priority this proposal would impact
  - Specify which strategic sub-priority this proposal would impact (if applicable)
  - Validation from priority advocate (or other person who owns strategy road map)

- **Advantages of using a Word/Excel template for portfolio intake:**
  - In most organizations, people are familiar with Microsoft Office including Word/Excel and have no problem with filling out templates composed in this software
  - The familiarity of users ensures that user adoption is high and does not become a factor by people not using the intake process
  - The familiarity of users with the native software of the templates ensures that little to no user training is needed

- **Challenges of using a Word/Excel template for portfolio intake:**
  - While Word/Excel are familiar to most people, this may lead to version control issues and proliferation, with people making their own modifications to the template

- Since the information is in separate files, it is not easy to aggregate or query the data to obtain insights or trend information
- While Word/Excel templates are easy to start with, experience shows that they become harder to manage with volume increases as time goes by

BUILDING AN INTAKE SOLUTION USING SHAREPOINT

The other recommended solution is to use SharePoint to create an intake form. Some basic SharePoint skills and the SharePoint platform itself are needed. The advantages of this approach are:

- Real-time visibility for users—promotes user satisfaction
- Web-based form is easy to use and can guide user data entry
- Easy to filter and create views
- Easy to create workflows
- Able to query aggregate data
- Useful for historical storage and retrieval

The challenges of using SharePoint to build an intake include some training and change management.

SUCCESS FACTORS FOR PORTFOLIO INTAKE

1. Identify Portfolio Type and Design Intake Accordingly

The intake design has to match the type of portfolio. A functional portfolio will need a speedy intake or it will choke on the volume of proposals. A strategic portfolio will need a carefully filtered intake or it will let through nonstrategic proposals which undermine the portfolio's functioning. A hybrid portfolio will need a well-crafted intake that is able to handle both types of proposals with the appropriate velocity and oversight. Identifying the type of portfolio is the critical first step.

2. Establish an SLA for Processing Proposals and Strive to Meet It

A portfolio can establish significant credibility by defining an SLA for processing a proposal and consistently meeting that SLA. What does this mean in real terms? For example, a portfolio office can commit to providing a decision on a proposal within one week of receiving it. For a strategic portfolio, this may take longer and needs a longer SLA. It needs to be noted here that the SLA should be declared after a careful study of all the intervening steps from receipt of the proposal to the disposition decision. Once declared, every effort should be made to meet the SLA and demonstrate to the organization that the SLA is being met.

One way of meeting the SLA is to publish a report at regular intervals showing how many proposals came through the intake process for that period and how many were approved (or denied) within the time frame of the SLA. While this can be a little uncomfortable in the beginning (most intake processes take a little bit of time to find their stride and start processing proposals—during that time, the SLA may not be met in all instances), it offers a huge incentive to identify and remediate the instances when the SLA was not met. It also serves a dual purpose of showing to the organization that the portfolio intake is performing well and underscoring the point that all proposals need to start using the process.

3. Expect to Do Plenty of Handholding

Stakeholders will need help filling out the project proposal template until the whole process gains traction. The portfolio office should be willing to help answer questions regarding the template and the whole intake process. If the portfolio office and staff are not approachable, stakeholders are likely to resist the process because they don't understand it.

4. Maintain a Web Page to Keep Track of All Proposals

One widely noticed pattern in most portfolio intakes is that there soon tends to be a mass of intake proposals, which is hard to manage without a specific mechanism. This mass of proposals, most of them at different stages of processing, creates a logjam that is typically frustrating for stakeholders to navigate through. To avoid this situation, the portfolio manager needs to implement a simple system of putting all the available information on a web page on the intranet that people can look up on their own and understand where their proposal is in terms of processing.

5. Simplify and Overcommunicate

A fair amount of confusion can be expected when intake is first adopted. It is recommended that the whole process be as simple as possible, and that the portfolio office compensate for the newness of the process by overcommunicating. As the organization becomes familiar with how intake works, additional complexity can be slowly introduced.

6. Take Political Cover

Dissatisfied stakeholders may use a slow intake process as an opportunity to criticize the portfolio office. It is therefore recommended for the portfolio office to always *take cover* by communicating clearly and demonstrating in writing that all possible help was offered to stakeholders concerning the intake process. Other defensive tactics include publishing an in-process list of proposals at every step and showing that stakeholders were regularly informed about the status of their proposals.

LEVELS OF PORTFOLIO CAPABILITY MATURITY

Level 1

- No defined intake process—there isn't a standard way as to how projects are identified and started.
- No distinction between functional and strategic proposals—proposals are handled as they are submitted, often slowing down the pace of processing.
- Due to a lack of an organized intake, the organization may be frustrated with the lack of speed in processing and also the lack of predictable, standard steps.
- Little to no governance oversight over starting projects—as a result there isn't a regular pathway to initiate projects.
- There are no standard intake criteria to qualify a proposal—every project proposal is considered case-by-case for approval and funding.
- Since there is not a defined body that provides approval to start projects, some projects get started soon and others may stagnate.
- There isn't a clear idea of the impact to portfolio funds from approving a project. Consequently, too many or too few projects may be approved, causing a potential overspend/underutilization situation at year end.
- There is little to no validation performed on projects that are purported to be strategic. There may not be a standard strategic plan to compare each project against in order to arrive at a decision.
- There may not be an annual planning process in place. This, in turn, prevents the creation of a queue of vetted projects that can quickly be considered for initiation.

Level 2

- A formal intake process exists and there is awareness that all projects have to go through intake. However, the intake process may not be optimized.

- Although a conceptual distinction exists between functional and strategic proposals, there may not be separate processing pathways for strategic and functional projects.
- A lack of separate processing pathways for strategic and functional projects results in some confusion about what steps apply to which type of proposal—causing proposals to take longer to process than expected.
- Although some governance exists before initiating new projects, there may not be sufficient governance in place before committing to strategic proposals.
- Although there are standard intake criteria to qualify a proposal, the criteria may not be comprehensive.
- A lack of comprehensive criteria may introduce some variability in assessing project proposals, leading to some projects getting approved quickly, while others may take longer.
- As part of intake assessment, there is some analysis done about the impact to portfolio funds resulting from approving a project. Consequently, the portfolio avoids approving too many or too few projects.
- There is some validation performed on projects that are purported to be strategic. However, there may not be an officially approved strategic plan to compare each project against in order to arrive at a decision.
- Although there may be an annual planning process in place, it may not be functioning at a level that can generate a reliable *queue* that forms a steady source for intake. Consequently, project proposals are more ad hoc in their genesis.

Level 3

- A formal, optimized intake process exists and there is awareness that all projects have to go through intake.
- Separate pathways exist for functional and strategic proposals—both are optimized for their respective type in terms of throughput and oversight.
- Having well-defined, separate processing pathways for strategic and functional projects allows proposals to be accelerated through intake without losing quality.
- Adequate governance exists for new projects, especially for strategic big bets.
- There are comprehensive intake criteria to qualify a proposal, ensuring that all proposals are assessed in a standard manner.
- As part of intake assessment, sufficient analysis is performed about the impact to portfolio funds resulting from approving a project. Consequently, the portfolio avoids approving too many or too few projects and optimizes the throughput of the portfolio.

- Comprehensive validation is performed on projects that are purported to be strategic. Strategy management is advanced and contains all the roles and artifacts, such as officially approved strategic plan and priority advocates, which enables the validation of each proposal to ensure strategic fit.
- There is a well-functioning annual planning process in place, along with a *queue* of projects that forms a steady source for intake.

CHAPTER SUMMARY

In this chapter, we began by stating the importance of portfolio intake in ensuring that portfolio execution is focused on meaningful items of strategic importance and financial viability. The chapter then explored the characteristics of the ideal financial portfolio intake in order to arrive at the characteristics of the ideal project portfolio intake. Next, the chapter covered the three broad classifications of portfolios and the needs imposed on the portfolio intake by the corresponding type of portfolio. The chapter then explored the actual workings of the strategic and tactical portfolio intakes. Next, the typical challenges in managing the intake process were considered, and the role of annual planning as a complementary function to intake was explained. Finally, best practices on implementing and managing portfolio intake processes were explained before concluding the chapter with the levels of maturity in this portfolio capability.

3

PORTFOLIO ANNUAL PLANNING

INTRODUCTION

Portfolio annual planning is the process of deciding, as an organization, which programs and projects to fund and execute in the next year. In some places, this might involve deciding what to do for the next two or three years. During this process, the whole organization is made to articulate all of the collective demand of the enterprise and put it all on the planning table to be prioritized in its entirety. In this chapter, we will cover the following topics related to annual planning:

1. State the importance of annual planning and explore the different ways in which it is beneficial to portfolio management
2. Explore the execution of annual planning in three distinct iterations or *passes*
3. Explain how annual planning data is gathered in Pass One along with an introduction to the annual planning template and discuss the nuances of populating it correctly
4. Explain how the collected data is aggregated, compiled, and displayed in Pass Two along with instructions to create the annual planning materials package
5. Explain how to orchestrate Pass Three of annual planning
6. List out the key outputs of the annual planning exercise
7. Explore pointers to success in annual planning

THE IMPORTANCE OF ANNUAL PLANNING

Annual planning is one of the foundational requirements of a high-performing portfolio. Although this exercise is complex and time-consuming, there are numerous benefits that flow from a well-executed annual planning process. Several of the benefits are related to each other and combine to deliver a comprehensive boost to the organization. These benefits are explored in the following section.

Discovery

In a large organization, information tends to fragment and distort; therefore, not everyone is aware of what other departments or teams are doing or plan on doing. Through the annual planning process, everybody is made aware of the work that other departments or teams are planning to do in the upcoming year. This discovery is valuable in several ways—as can be seen in the upcoming sections.

Redundancy Elimination

Redundancy elimination is one of the big advantages of discovery. The organization can spot and eliminate redundant efforts, such as two departments that are both trying to implement the same kind of system. These two departments could combine their efforts, or one department could go do something else, or at the very least, they could negotiate a volume discount from a potential vendor.

Another type of redundancy that can be discovered and eliminated refers to systems that may be in the process of becoming obsolete. For example, consider a department that is planning to do a project to write a series of reports for a particular system. As part of annual planning, it may be discovered that this system is going to be discontinued, thus, it is no longer feasible or necessary to write reports for that system. The department could therefore cancel the report-writing project and redirect the resources elsewhere.

Sequencing

The next big advantage of discovering all the demand in the organization is that it enables the sequencing of complex efforts. When dealing with complex projects/programs, there is often an optimal sequence to be followed in terms of which components to implement first. Furthermore, there are always constraints of budget, manpower, and bandwidth that prevent everything from being done at once. By knowing the collective demand of the organization, it is possible to arrive at this logical sequence of subcomponents that need to be taken up for completion within the constraints of resources.

Green Lighting

Green lighting is defined as the accelerated approval of pre-vetted proposals known to be of importance. In a large portfolio, it is likely that a large number of proposals will be seen during the course of the year, and it is inefficient to perform a deep dive on each of them when they present themselves. By reviewing the bulk of the organization's demand through an annual planning process, the projects that are vital to the company's strategic vision can be earmarked ahead of time and green lighted when they are ready to execute.

Populating the Queue

Another advantage of the annual planning process color is to maintain a queue. In a portfolio, there is always a need for keeping a list of viable proposals that can absorb any dollars that may get freed up due to underspend or termination of other projects. In the annual planning process, we arrive at a comprehensive list of projects that the company wants to do and can be matched to the funding that may become available later on in the year.

Strategic Planning

Perhaps the biggest advantage of the annual planning process is that it reinforces strategic planning and the direction of the organization in executing to the strategic plan. How does it do that? Quite simply, by applying an overlay of the strategic road map on top of the combined demand of the organization to identify which proposals are in line with the strategic direction and which are not really aligned to the company's stated strategy. This is also a great time to revisit the progress made on the strategic road map and benchmark the progress made thereof. This is covered in greater depth in the next section.

THE EXECUTION OF ANNUAL PLANNING

At a high level, annual planning is done in three distinct iterations or *passes*. It is useful to compartmentalize the activity into the following iterations, as it helps the organization keep track of the current status in this months-long effort.

- *Pass One—demand gathering*: All the project owners participating in annual planning need to fill out a standard template. This is done to ensure that all the demand is captured and qualified in identical terms.

- *Pass Two—demand analysis*: All the captured project demand data is analyzed and presented in views that enable governance body/decision makers to make an informed decision in ranking the priority of the projects.
- *Pass Three—matching demand to supply*: The total demand of the prioritized list of projects is compared with available funds and the *cutoff line is drawn*—projects above this line are the projects that are funded for the next year. The rest of the project demand is preserved in a queue so that if additional funds become available, they can be taken up for execution.

PASS ONE OF ANNUAL PLANNING

As mentioned before, Pass One consists of gathering all the project demand in a standard template (the components of template are explained in detail in the next section of this chapter). Here are the high-level steps in Pass One, as depicted in Figure 3.1:

1. The portfolio office typically starts Pass One by holding a kickoff meeting with all the department representatives and informs them that the annual planning is being kicked off.
2. Then the template is shared with the group and the important elements of the template are described so that the group understands what information is being sought. The departments discuss what proposals to put forth as part of their demand.
3. The stakeholders fill out the templates accordingly and submit back to the portfolio office.
4. The portfolio office reviews the submitted templates.

Key Components of the Planning Template

The planning template is the key document in the whole annual planning process. Here is a walkthrough of the key sections found in the template:

- **Project Description**: A succinct description of what the project is. If the project belongs to a program, this would be the place to note it.
- **Project Strategic Component**: The contribution of the project to the strategic road map is described here. This is an important piece of information that needs to be sourced carefully in order to make optimal decisions in selection and funding. (Please see a later section called: How to Approach the Project Strategic Component Field and Ensure that It Is Filled Out Correctly.)

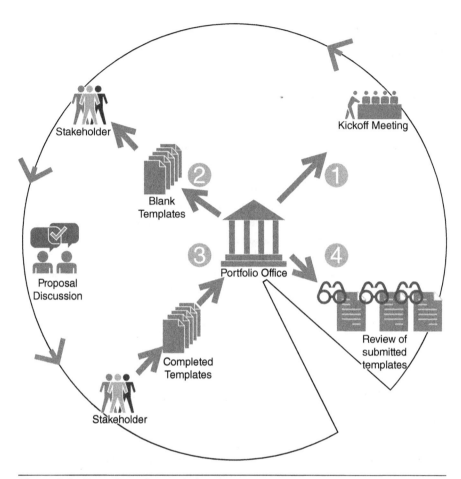

Figure 3.1   Pass One of annual planning

- **Project Financials**: Demand and return on investment (ROI): This section describes the total demand of the project and the cost components of that demand. It also captures the ROI.
- **Project Benefits**: This section spells out the benefits expected to accrue by executing the project.

The next section contains a detailed description of each of the previously mentioned elements. It is vital to capture these as accurately as possible to have an annual planning process that actually delivers value to the organization.

How to Ensure that the Project Description Is Filled Out Correctly

Figure 3.2 shows the portion of the annual planning template that contains the project description and associated fields. Providing guidance and examples that show the project owners how to fill out the project description is recommended (see best practices at the end of Chapter 3 about holding an annual planning workshop to walk stakeholders through the templates). Additional pointers to make this field effective are as follows:

- Ensure the standard name of the project is used and that this does not change. While it sounds obvious, experience during annual planning has shown a lot of avoidable confusion created by project names being inconsistent during the various iterations. (Example: Project Office On-line → Project Office Cloud → Project Office 365 → Project O365, etc.)
- The project description should be jargon free and coherent to a decision maker who may be removed from the domain and details.
- If this is a continuing project, it should be called out as such and a financial snapshot provided in the financials' section.
- If this project is part of a program, that fact needs to be validated by the program manager and the name of the program noted.

How to Approach the Project Strategic Component Field and Ensure that It Is Filled Out Correctly

Beyond the financial estimates and ROI projections, one of the trickier aspects of demand collection is answering the question of "What does this proposal add to our strategic road map?"

Annual Planning Template						
				Hard Benefits	Soft Benefits	Benefits Validated?
Project Name		New/Continuing	NPV			☐ **Finance**
Project Contact		Start Date	ROI			☐ **Portfolio Office**
Strategic Priority		End Date	TCO			
Strategic Sub Priority						

Figure 3.2   Basic information section of the annual planning template (NPV = net present value, TCO = total cost of ownership)

This is hard because of the following factors:

- Project owners are typically conversant only with the details of their specific project, not the *big strategic picture* or where this fits into that big strategic picture.
- Project owners sometimes try to get their project a better chance of funding by aligning with whatever seems to be the key strategic theme *du jour*.

In most organizations, there is often confusion about what exactly the strategic road map is and how the rank and file can relate to it. So even well-meaning project owners may not be able to comply with the directive of choosing which strategic aspect would be influenced by their project. How do we remedy this situation? Enter the role of priority advocate. The role of the priority advocate can be explained with the artifact called the strategic decomposition table, which was introduced in Chapter 2 and is reproduced in a slightly modified form in Table 3.1.

As seen in Table 3.1, the aspirational high-level strategy is broken out into various concrete strategic priorities. Each strategic priority typically has a multi-year life, with different strategic milestones along the way. Figure 3.3 shows the multi-year journey of strategic priority A. The yearly milestones of attainment for that strategic priority are shown along the top. Under each yearly milestone, all of the projects that are requesting funding as part of that year's annual planning are shown. The understanding is that all of the projects directly under a

Table 3.1 Modified form of a strategic decomposition table

High-Level Strategy	IT Strategic Priority	Description of Strategic Priority	Sub-Projects Aligned to the Strategic Priority
Aspirational: Strategic Vision for the Future	1	Description of Strategic Priority #1	Sub-project 1
			Sub-project 2
			Sub-project 3
	2	Description of Strategic Priority #2	Sub-project 4
			Sub-project 5
			Sub-project 6
	3	Description of Strategic Priority #3	Sub-project 7
			Sub-project 8
			Sub-project 9
	4	Description of Strategic Priority #4	Sub-project 10
			Sub-project 11
			Sub-project 12

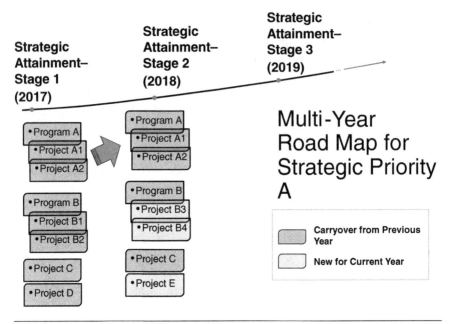

Figure 3.3   Multi-year strategic road map for strategic priority *A*

yearly milestone contribute toward the attainment of that particular milestone. The priority advocate builds and maintains the entire road map for each strategic priority.

The following are the roles and responsibilities of the priority advocate:

a) The priority advocate owns the strategic road map and has been designated as the point person for that strategic priority of the organization's strategic road map. (Refer to Chapter 1 to review how the different dimensions fit into the overall strategic road map.)

b) The priority advocate engages with the project owners and analyzes how/whether their project fits with that particular strategic priority (hence the term *priority advocate*). For example, in Figure 3.3, Program A has been found to be a valid fit under strategic priority A. There may be other projects, G, H, or F that were analyzed and found not to be a valid fit—hence, they are not shown in the strategic priority's road map.

c) The priority advocate validates the strategic impact of a project proposal, as documented in the annual planning template, and assigns a relative score if necessary. (Example: Project A1 has a relative score of nine versus Project C's score of seven as it relates to influencing the strategic dimension. Alternatively, strategic impact [to the same specific dimension] can be described as high/medium/low).

d) The priority advocate updates the strategic road map with new projects that fit within the priority and get funded as part of the annual planning.

With the help of the priority advocate, here's how the process unfolds to fill out the annual planning template for the strategic section:

- Project owner approaches strategic priority advocate with their project details
- Priority advocate reviews strategic fit within their strategic priority
- If project is a fit within that priority, priority advocate provides guidance to choose right drop-down choice in template (two fields in the template are used for this purpose: strategic priority and strategic sub-priority)
- Priority advocate also assigns a score for extent of influence on strategic priority—this score will be useful later, to create an overall priority score for the proposals
- Project owner checks the *priority advocate reviewed* flag and submits the planning template
- Ensures that the strategic dimension is a choice among a standard menu of choices—not a free-form field
- Ensures that this field has a check box for priority advocate review and approval

More details about the strategic planning process are found in Pass Two of the annual planning process.

How to Ensure Project Financials Are Captured Correctly

Project financials refers to project benefits and project costs. This section explores how to fill out the financials correctly for any project. Figure 3.4 shows the portion of the annual planning template that contains the project benefits. Project benefits span a wide spectrum. There is a detailed explanation of the different types of benefits and the qualification for each type in Chapter 7. To ensure uniformity, the annual planning template should reflect the same benefit categories that are used in the benefits realization process as detailed in Chapter 7.

Next, the cost categories need to be captured adequately. Figure 3.5 shows the portion of the annual planning template that contains the project benefits. Here are the typical cost categories in a project:

- One-time/Project Implementation Costs
  - IT OPEX cost
  - IT CAPEX cost

Project Financials–In Thousands
Benefits

	IT/Business	Category (Hard/Soft)	Op/Cap	2017	2018	2019	2020	2021	Total
Enter full $ amounts. Formatted as thousands	IT	Hard - P&L Impact	Op						0
	IT	Hard - P&L Impact	Cap						0
	IT	Hard - Non P&L Impact	Op						0
	IT	Hard - Non P&L Impact	Cap						0
	IT	Soft Benefit	Op						0
	IT	Soft Benefit	Cap						0
	Business	Hard - P&L Impact	Op						0
	Business	Hard - P&L Impact	Cap						0
	Business	Hard - Non P&L Impact	Op						0
	Business	Hard - Non P&L Impact	Cap						0
	Business	Soft Benefit	Op						0
	Business	Soft Benefit	Cap						0
Total Benefits		Hard Benefit Total		0	0	0	0	0	0
		Hard + Soft Benefits Total		0	0	0	0	0	0

Figure 3.4   Project benefits section of the annual planning template

Cost

	Actual								
	IT/Business	Category	Op/Cap	2017	2018	2019	2020	2021	Total
Enter full $ amounts. Formatted as thousands	IT	Investment	Op						0
	IT	Investment	Cap						0
	IT	Ongoing	Op						0
	IT	Ongoing	Cap						0
	Business	Investment	Op						0
	Business	Investment	Cap						0
	Business	Ongoing	Op						0
	Business	Ongoing	Cap						0
Total Cost		IT Investment Subtotal		0	0	0	0	0	0
		Bus Investment Subtotal		0	0	0	0	0	0
	IT + Business Ongoing Support Subtotal			0	0	0	0	0	0
		Grand Total Cost		0	0	0	0	0	0

Figure 3.5   Project costs section of the annual planning template

- • Business OPEX cost
- • Business CAPEX cost
- • On-going/Operational Costs
  - • IT OPEX cost
  - • IT CAPEX cost
  - • Business OPEX cost
  - • Business CAPEX cost

In the case of multi-year projects, costs and benefits need to be shown in the correct year, for transparency.

Note that all of the procedures for capturing information were based on using an Excel template. Conversely, a web-based form could be used (more about this in Pass Two). The reason we focused in detail on an Excel file is that most organizations feel comfortable e-mailing an Excel file back and forth as they go through iterations to arrive at the correct annual planning details of a project.

PASS TWO OF ANNUAL PLANNING

As we discussed before, Pass Two of annual planning consists of aggregating all of the demand that was gathered in Pass One and then presenting it to the decision makers. This can be challenging because there is a need to sift through a considerable amount of material and present the same to the executive decision makers in a concise manner—all while preserving the complexities inherent in the demand. Figure 3.6 shows the complete orchestration of Pass Two at a high level. The details are described in the following section.

Execution of Pass Two

1. Aggregating the Gathered Information

The excel templates described in the previous section are user-friendly when it comes to the project owners having to fill out the information and e-mail the files to each other, but they are really not easy for either the portfolio manager or the decision makers to work with. Therefore, this information will have to be aggregated to a format that is more amenable to presentation and analysis.

These two opposing design principles—user-friendly versus efficiency—need to be reconciled in order to proceed with Pass Two. Here are some options to accomplish the same:

- • The Excel templates can be built using VBA or InfoPath (or any other smart field setup) such that they can programmatically retrieve all the information in the worksheets (the services of a competent VBA/Excel programmer would be needed to create this setup).

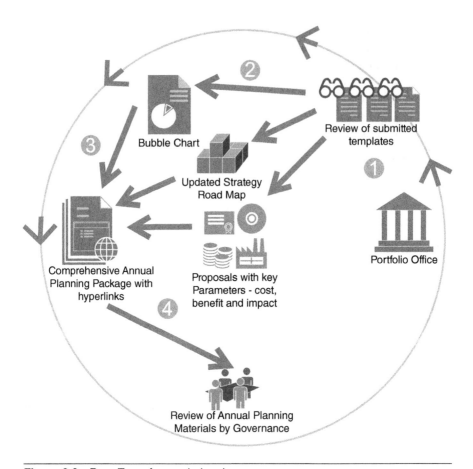

Figure 3.6   Pass Two of annual planning

- The Excel template can be integrated with SharePoint features that have enhanced interoperability with MS Office (you would need a competent SharePoint programmer to create the setup).
- Brute force: Have a team of interns or temps reenter the data in Excel into a database or other aggregate list that can then be queried and reported for analysis. While this is effort intensive and painstaking to perform, this may be the most expedient option for many organizations that need to get going with annual planning.
- The other option, as explained in Pass One, is to use a web-based form to collect all the details for annual planning for each project. This is orders of magnitude more efficient and easier to aggregate compared to an Excel sheet because the information from a web-based form is written to

a database immediately upon submission, making it straightforward to create reports for analysis.

2. Displaying the Aggregated Information

Annual planning yields a huge amount of data. It is not easy for an executive decision maker to assimilate all the data, even if it is provided in a list form that can be sorted. Some kind of visualization tool is needed to crunch the data into a view that can be easily grasped and decided upon. A useful visualization tool in this space is the *bubble chart*. A bubble chart is a variation of a scatter chart in which the data points are replaced with bubbles, and an additional dimension of the data is represented in the size of the bubbles. (For example: consider two projects that are identical in every way but cost—the project with a larger cost will appear as a bigger bubble than the one with a smaller cost.)

Figure 3.7 shows an example of a bubble chart. This example chart shows cost versus profit versus strategic value. In this particular example, strategic value and cost form the two axes, while the size of the bubble shows the profit; however, this can be configured to make the bubble size show any attribute—such as cost or strategic value. Bubble charts can be created using either Microsoft Excel or a dedicated software such as Bubble Chart Pro.

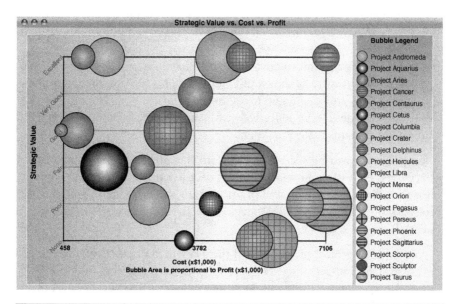

Figure 3.7   Bubble chart showing all projects gathered through annual planning (Image courtesy of Bubble Chart Pro)

Another artifact that needs to be updated with the collected annual planning data is the strategy road map (shown in Figure 3.3). It helps to visualize how the new demand lines up with the existing strategic road map.

3. Creating the Annual Planning Materials Package

The key to the success of annual planning is getting the materials to the decision makers in a comprehensive but user-friendly package. This package is typically a large PowerPoint (PPT) accompanied by optional Excel files holding the raw data. The steps to create this package are outlined here:

- First, the annual planning demand is aggregated into a list using one of the methods outlined in the aggregation section (*1. Aggregating the Gathered Information*).
- This list is used to create a bubble chart as described in the representing information section (*2. Displaying the Aggregated Information*). The bubble chart is then stored in the annual planning PPT.
- In addition to the bubble chart, decision makers find it useful to receive an actual concise list of the projects along with one or two of the most important attributes (cost and benefit come to mind immediately). This concise list is created and is included in the annual planning PPT.
- It's also useful to create a one- or two-page summary for each project and keep this in the appendix of the annual planning PPT. The concise list can have hyperlinks that jump to the corresponding summary in the appendix.
- Finally, it's also informative to include the multi-year road map for each strategic priority as shown in Figure 3.3. This strategy road map shows how each project lines up to a strategic priority.
- These materials are collectively circulated to the decision-making group to help them create a ranking.

A key point to ensuring success in Pass Two is to ensure that the data must be as accurate as possible to enable correct decision making. When it comes to measuring strategic fit, the input of priority advocates is vital in ensuring that projects score accurately on the strategic scale. In summary, here are the outputs of Pass Two:

- A list of project proposals and their key parameters: cost, benefit, and strategic impact
- A representation of where the projects stand with respect to each other— also known as a prioritized list with key stakeholders providing the ranking
- All of the above are compiled into a comprehensive package called the annual planning PPT

PASS THREE OF ANNUAL PLANNING

In Pass Three we're made aware of how much budget exists for the portfolio in the new year. The challenge then is to maximize the value delivered to the organization within the constraint of this budget number. In most organizations, the combined project demand generated by the annual planning exercise will exceed the available funds. Since the list of projects has been prioritized in Pass Two, it is a fairly straightforward exercise to decrement the available funds according to the demand of the top-ranked projects and then *draw the line* when the money runs out. Figure 3.8 illustrates the high-level process of

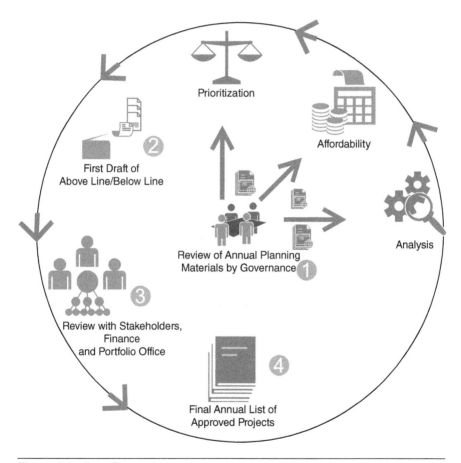

Prioritization

Affordability

First Draft of
Above Line/Below Line

Analysis

Review of Annual Planning
Materials by Governance

Review with Stakeholders,
Finance
and Portfolio Office

Final Annual List of
Approved Projects

Figure 3.8 Pass Three of annual planning

how supply and demand are matched in Pass Three of annual planning. The details of this process are explained here:

- The decision-making body reviews the available funds along with the demand of the top-ranked projects and determines what's *above the line* (funded) versus *below the line* (unfunded).
- The implications of not funding certain projects (also known as the *cost of not doing*) is shared with the organization and other stakeholders to determine if they are creating an unacceptable risk to the organization with the present *above/below* line proposal.
- Typically, some relief is obtained in the form of additional funds, which make it possible to squeeze in a few more projects above the line. Alternatively, the organization may decide that funds need to be carved out of the existing *above the line* projects in order to provide for a few *below the line* projects.
- After some jostling and rearranging, the organization settles down with a *best-possible* basket of projects to go forward with. It's a great idea to update Figure 3.3 to show *funded versus nonfunded* projects so that governance and decision makers are aware of the impact to the strategic road map.
- The *below the line* projects are turned into a queue that can now absorb any new funds that may become available during the year.

KEY OUTPUTS OF THE ANNUAL PLANNING EXERCISE

- **Ranked list of projects**: A list of projects ranked according to importance—the higher ranking projects are funded and form the annual plan
- **Annual Plan**: A list of projects that have been funded in the next year
- **Updated Strategic Road Map**: Validation of strategic direction and prediction of how the chosen projects will influence progress toward strategic destination
- **ROI of each project**: Cost and benefits, along with classification of benefits
- **Project description snapshot**: A brief explanation of what each project does

POINTERS TO SUCCESS IN ANNUAL PLANNING

1. Launch Annual Planning Exercise with Comprehensive Workshop/Training

The success of the annual planning exercise is closely dependent upon meaningful participation by the organization. Participation in annual planning, in turn, is greatly facilitated by a comprehensive workshop that addresses all the questions surrounding annual planning. This workshop is also an appropriate venue to share and demonstrate how to use the various templates and other artifacts for the annual planning exercise. The workshop should also cover the timeline of the annual planning exercise. It's even more effective to archive this training session, along with relevant resources for later reference and use by the organization. A mind map may be the right tool to help the rank and file navigate the possibly complex scenarios involved in annual planning. Chapter 20 has an introduction to mind maps and how to use them.

2. Follow Consistent Treatment for Existing Projects

Few portfolios start from a blank slate during annual planning. What should be done with existing projects? Should existing projects be prioritized along with new projects? Or should they be kept whole (funded for their full demand for the next year) and the remaining funds shared among new projects? Here are some recommendations:

- Existing projects should be included in the exercise, but should be given a higher priority compared to new projects because of the money already invested in them—but only as long as the existing projects are performing well. If an existing project is not performing well, it should not be automatically elevated over new projects.
- If an existing project is significantly underperforming, annual planning could provide a venue to kill such underperforming projects—therefore, we may need to look at modified earned value management (*mEVM*) data for existing projects to support a decision on discontinuing them (see Part 2 on how to do mEVM).

3. Navigate Annual Planning with an Existing Enterprise Portfolio Management Tool

This whole chapter describes how to perform annual planning entirely outside of an enterprise portfolio management tool. But how does annual planning take place when the organization already has in place an established enterprise portfolio management tool? Based on the maturity of the organization, two choices are possible:

- If the organization is comfortable working within the tool, consider using the scenario planning features of the tool. This will work for only a few organizations since it requires a high degree of portfolio maturity.
- The more popular experience shows that the rank and file at most companies are more comfortable working within MS Office—specifically Excel and PPT. In this case, orchestrate all the annual planning outside of the portfolio tool (as explained in this chapter) and enter into the portfolio tool when annual planning outputs are final (see earlier section for annual planning outputs).

4. Guard against Bad Data and Hold People to Promises

The biggest problem seen during annual planning is that people will submit project data that is wildly inaccurate (imagine completely incorrect timeline, budget, and benefits). In their defense, the project owners would say, "It's too far in the future to predict accurately. This is my best guess for now. Fund me so I can submit accurate estimates."

Some organizations use a confidence level (ranging from 0% to 100%, with 100% signifying total certainty in the estimate) to accompany the cost and benefit estimates. However, this has yet to be proven to actually add any value. In fact, this becomes a convenient loophole for many people who will submit estimates with low confidence levels and then justify the resulting variance using the same confidence levels. Others game the system with 50% estimates—high enough to seem credible and low enough to justify the variance when they miss the mark. Here are some alternative approaches to increase data integrity for better planning outcomes:

- **Promote integrity in benefit data**: Projects that promise hard benefits should be given first preference in funding. However, each project that promises hard benefits should have a corresponding arrangement worked out with the finance department where there are financial consequences for the hard benefits promised. (Example: a project that promises to save $1M in actual spend will need to have the budget reduced by a corresponding amount). This is a high bar to clear, and only the most confident projects with actual hard benefits will step forward—which works, since we only have limited funds and need to first focus on the projects that actually deliver a return. Another effective way to assure benefits materialize as promised is to engage the benefits review process, as explained in Chapter 7.

- **Promote integrity in demand data**: Projects that have a large demand for funds run a risk of not using the funds as planned and then return it too late in the year, effectively blocking other projects. It's always a good idea to engage the mEVM process, which will accurately provide early warning about underspend concerns. Another good practice is to allocate only a fraction of the total ask for all projects with the understanding that some will underspend. In any case, follow *portfolio funding best practices* as explained in Chapter 4.
- **Provide discovery funding only**: Some projects seem to be good bets but simply have no data to substantiate that conclusion. In such cases, the recommended approach is to fund the project enough to come back with solid estimates—this is assuming that the project is recognized as a critically important priority.

5. Preserving Annual Planning Continuity by Avoiding a Total Reboot Each Year

One of the biggest complaints about annual planning is that the organization seems to do a complete reset each year—learning and using nothing from similar massive efforts in past years. It has frequently been compared to *boiling the ocean*—every year.

However, it may not be really necessary to do *a complete new boil* each year. A well-crafted process will allow for a more modest and easier effort each year—by only managing the changes from year to year. In other words, strategic demand does not change dramatically from one year to the next, and it may be enough to just track the following *puts and takes* each year:

- Projects whose demand numbers changed from last year
- Projects from last year that are no longer viable proposals
- New proposals that were not seen last year

A huge factor in preserving continuity is to not change the templates and approach each year. Adequate change management should be in place and emphasize that the change each year is minimal and only what's really necessary.

6. Ensure Follow Through for the Annual Planning Exercise

Another frequent criticism of the annual planning process is that little if any follow through is seen regarding the artifacts seen during the process. In short, after the annual planning period finishes, things *go back to how they were* and the whole exercise seems to be for nothing. This seems especially true for projects that did not get funding.

The only remedy for this situation is a conscious decision on the part of the portfolio manager and the stakeholders to keep revisiting the carefully made annual plan and use it as a key input to drive portfolio decisions. For example, a well-made annual plan will have a queue of solid project proposals that have been vetted and are awaiting funding; these projects should be the first choice in which to invest any funds that become available during the year. At every possible opportunity the annual plan should be demonstrated as the *master plan* that drives portfolio decisions. This serves to underscore the importance of annual planning to the rest of the organization and, in turn, drives increased participation.

7. Ensure a Link between Strategy and Annual Planning

One of the characteristics of annual planning is that it shines a spotlight on the performance (or absence!) of the strategic planning function in the enterprise. A strategic planning process that is not grounded in the actual direction or road map of how the enterprise plans to transform itself will be found wanting when the time comes to map project proposals to it.

The remedy for this situation lies in recognizing that effective strategic planning needs to be in place for annual planning and also for portfolio management to work. Well before the annual planning cycle starts, the organization needs to invest in the process and socialization that is necessary to produce a stable strategic plan that extends at least for the next few years. Having a well-socialized plan in place along with advocates for each strategic priority will vastly simplify the annual planning process and succeed in matching proposals to key drivers of strategy.

8. Avoid Making Concentrated Bets

The following scenario is typical of annual planning instances at most organizations:

- Imagine Project Proposals A through Z participate in an annual planning exercise
- Projects A and B promise a lot of benefits, have significant asks in terms of allocation, and take up the lion's share of the available portfolio budget
- Midway during the year, it's discovered that the budget demand for A and B is vastly overstated, and these two projects do not need the stated amount of funds

- The portfolio tries to reallocate the money, but there isn't much time left in the year to execute other projects
- The year ends with significant leftover funds and few wins (assuming A and B haven't overstated the project benefits too)

The following is the recommended method of avoiding the previously described scenario by spreading the bets around:

- Project Proposals A through Z come to an annual planning exercise
- Projects A and B promise a lot of benefits and have significant asks in terms of allocation
- The portfolio manager provides A and B with a fraction of the funds demanded and provides the rest to projects C, D, and E
- All projects follow mEVM governance (as explained in Chapter 5)
- Projects A and B spend less than what they forecasted, the rest of the projects spend as planned, and the year ends with little funds left over as well as solid strategic gains from the execution of five projects (A through E) as compared to two projects in the previous scenario (just A and B)

9. Preserve Every Year's Annual Planning Submissions for Reuse Next Year

Every annual planning exercise concludes with a sizable number of projects that did not get funded. These projects either need to try to get funding during the year or come back at the next annual planning exercise looking for funds. A significant number of projects do have to come back at the next year's planning session. For those projects, it would be a much simpler effort if the various departments could access their previous year's submission (for any particular project), make some edits and updates, and then resubmit the same project for consideration at this year's annual planning exercise. To enable this scenario, the project office needs to make an active effort to capture all the submissions from the annual planning exercise and then make it available in an *easy-to-navigate* manner for all the stakeholders. A recommended solution is to use the mind map technique, which allows for a concise representation of an array of topics and subtopics. Under the topic of annual planning, there would a subtopic containing links to the previous years' submissions, called *historical annual planning submissions*. This arrangement is shown in Figure 3.9. For additional information, please refer to Chapter 20, which contains an extensive discussion of mind maps and how to deploy mind maps to display portfolio content.

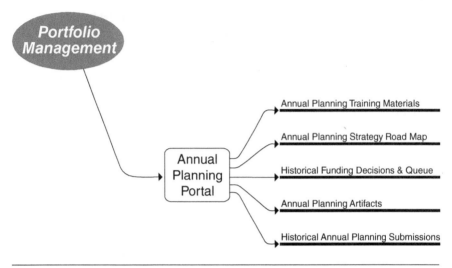

Figure 3.9   Mind map showing link to previous years' annual planning submissions

LEVELS OF PORTFOLIO CAPABILITY

Level 1

- The organization is either unfamiliar with annual planning or unconvinced about the need for annual planning. Consequently, there may not be a comprehensive exercise that encompasses the whole organization.
- Some kind of limited planning exercise may happen, but it may not have a real cadence that is typical of annual planning. These limited planning exercises may not adhere to a fixed schedule each year.
- In the absence of a comprehensive annual planning exercise, fragmented exercises may be conducted at the department or sub-organizational level. Some of the efforts might be duplicated and redundant.
- Due to the different entities running each planning exercise, there is a lack of uniformity. Each unit or department performs planning in their own way and it is hard to aggregate the bodies of demand together to produce an enterprise-wide demand.
- The fragmented planning efforts coupled with the absence of an overarching strategic framework significantly hampers the strategic considerations involved in the exercise as listed here:
  - There may not be a standard strategy taxonomy which can be used by the different projects to describe what they offer in strategic terms.

- There may not be a multi-year strategy road map to plan against. Consequently, projects have no visibility to the time frame of implementation to which they can align.
- Due to the lack of a common strategic framework, it may not be possible to assess all projects uniformly for strategic fit.
- Formal priority advocates may not exist to function as gate-keepers to assess the fit of proposals to the strategic road map. Consequently, projects may be falsely aligned to strategic priorities that are not relevant to those projects.

- Due to the lack of a standard set of steps in annual planning, it may not be possible to define phases of activity that help the organization keep track of the overall timeline of the exercise.
- The lack of a standard process and standard templates makes it hard to preserve continuity from one year's planning to the next. For the same reason, using the previous year's planning output is made hard by the templates and process changing every time.
- There may not be a streamlined system for collecting demand. Demand collection may be a disjointed arrangement with most stakeholders e-mailing templates to the portfolio office but with others having additional problems surrounding version control.
- The absence of an effective annual planning setup results in the inability to create a *queue* of vetted projects that can serve as a ready source to project intake for the next year.

Level 2

- The organization is familiar with annual planning and is aware of the need for annual planning. Consequently, there is a comprehensive exercise that encompasses the whole organization.
- The annual planning exercise has a standard cadence and typically adheres to a fixed schedule each year.
- The annual planning exercise is largely uniform across departments and business units, with everyone using the same templates and adhering to the timeline. This eliminates duplication and redundancy in efforts.
- Due to all the departments and units following the same process, there is uniformity in the collection of demand. This makes it possible to aggregate the bodies of demand together to produce an enterprise-wide demand.
- Although the planning exercise is no longer fragmented, the absence of an overarching strategic framework could still hamper the strategic considerations involved in the exercise, as listed here:

- There may not be a standard strategy taxonomy that can be used by the different projects to describe what they offer in strategic terms.
- There may not be a stable, multi-year strategy road map to plan against. Consequently, projects have limited visibility to the time frame of implementation to which they can align.
- Due to the lack of a common strategic framework, it may not be possible to assess all projects uniformly for strategic fit.
- Formal priority advocates may not exist to function as gate-keepers to assess the fit of proposals to the strategic road map. Consequently, projects may be falsely aligned to strategic priorities that are not relevant to those projects.

- The standardization in the annual planning makes it possible to define phases of activity that help the organization keep track of the overall timeline of the exercise.
- Although the process is standard, the templates may change every year, making it hard to preserve continuity from one year's planning to the next. For the same reason, using the previous year's planning output is not an easy task.
- There may be some attempts at building a streamlined solution to gather demand. This solution could take the form of a simple SharePoint site or other similar structure where stakeholders could upload their data and be able to retrieve it as needed. This solution would also be capable of managing versions.
- Although not optimized, the presence of an annual planning setup allows the creation of a *queue* of vetted projects that can serve as a ready source to project intake for the next year.

Level 3

- The organization is familiar with annual planning and is well aware of the need for annual planning. Consequently, there is a comprehensive exercise that encompasses the whole organization.
- The annual planning exercise has a standard cadence and typically adheres to a fixed schedule each year.
- The annual planning exercise is largely uniform across departments and business units, with everyone using the same templates and adhering to the timeline. This eliminates duplication and redundancy in efforts.
- Due to all the departments and units following the same process, there is uniformity in the collection of demand. This makes it possible to aggregate the bodies of demand together to produce an enterprise-wide demand.

- The uniformity of the planning exercise coupled with the presence of an overarching strategic framework serves as a significant catalyst for the strategic considerations involved in the exercise as listed here:
  - There is a standard strategy taxonomy that can be used by the different projects to describe what they offer in strategic terms.
  - There is a stable, multi-year strategy road map to plan against. Consequently, projects have full visibility to the time frame of implementation to which they can align.
  - Due to the presence of a common strategic framework, it is possible to assess all projects uniformly for strategic fit.
  - Formal priority advocates exist and they function as gatekeepers to assess the fit of proposals to the strategic road map.
- The standardization in the annual planning makes it possible to define phases of activity that help the organization keep track of the overall timeline of the exercise.
- The templates are kept stable and unchanged each year, making it straightforward to preserve continuity from one year's planning to the next. For the same reason, using the previous year's planning output is possible and contributes to a low-effort annual planning exercise year after year.
- There is a streamlined solution in place to gather demand. This solution could be a customized SharePoint site or other similar structure where stakeholders could upload their data and be able to retrieve it as needed. This solution would also be capable of managing versions and also producing advanced reports and other dashboard information.
- The presence of an optimized annual planning setup allows the creation of a *queue* of vetted projects that can serve as a ready source to project intake for the next year.

CHAPTER SUMMARY

Annual planning is a complex and time-intensive exercise. If executed well, it is a great enabler for a high-performing portfolio. To do it right, it has to be easy to use and it must align well with the strategic road map of the enterprise. Annual planning also has to be structured correctly to help decision makers pick the right projects within the available sum of funds. In this chapter, we began by exploring the importance of annual planning, in some depth, and covered the different angles in which it promotes the functioning of the portfolio. We then detailed the three distinctive iterations through which annual planning is performed. As part of these iterations, we also explored, in depth, how the key artifacts of portfolio management function—such as the annual planning template. We covered data aggregation, compilation, and packaging practices and explained why these are integral to producing the right decisions from portfolio governance. We also listed the key outputs of annual planning and rounded out with best practices that have been proven to work in facilitating annual planning in various organizations. We concluded the chapter with the levels of portfolio capability in this dimension.

4

PORTFOLIO FUNDING STRATEGIES

INTRODUCTION

The essential function of a portfolio is to manage the funds allotted to it and produce optimal outcomes with those funds. However, the management of portfolio funds can be a complex exercise that needs the portfolio manager to balance several competing constraints. In this chapter we examine the following aspects related to portfolio funding:

1. Discussion of the annual funding cycle and the two types of investment dollars typically encountered in the funding cycle
2. The supply side challenges imposed by the annual funding cycle on the portfolio office's choice
3. The demand side challenges imposed by the annual funding cycle on the portfolio office's choice
4. A discussion of the best practices, including run-hot factor, allocation, release, and turnback features to mitigate both supply side and demand side constraints
5. A discussion of a couple of different portfolio funding models that are commonly found in organizations

THE PORTFOLIO ANNUAL FUNDING CYCLE

Before we get into the details of portfolio funding constraints, it's useful to consider how the portfolio funding cycle works. Typically, a portfolio's allocation of funds for any given year is decided during the annual planning activity that occurs during the previous year. As part of that activity, the finance department sets aside a sum of money that is divided into OPEX and CAPEX. What are OPEX and CAPEX? Here is a basic definition:

> An operating expense, operating expenditure, operational expense, operational expenditure (OPEX) is an ongoing cost for running a product, business, or system.[1] Its counterpart, a capital expenditure (CAPEX), is the cost of developing or providing nonconsumable parts for the product or system. For example, the purchase of a photocopier involves CAPEX, and the annual paper, toner, power, and maintenance costs represent OPEX.[2] For larger systems like businesses, OPEX may also include the cost of workers and facility expenses such as rent and utilities.

Therefore, the portfolio starts the year with an OPEX and CAPEX budget. The next task is to allocate these to various projects in the portfolio and manage their actual expenditure (commonly called *actuals*) to stay within the budget numbers. The ideal end of the portfolio cycle entails finishing the year with as little (or zero) unused budget.

THE CHALLENGES IN MANAGING THE PORTFOLIO FUNDING CYCLE

The constraints encountered in optimizing portfolio spending during the year can be divided into two categories—supply constraints and demand constraints. Supply constraints are the cause of supply challenges, which refers to the consequences of actions and decisions taken by the portfolio manager in supplying funds to projects. Demand constraints are the cause of demand challenges, which refers to actions and decisions on the part of the projects which have an impact on the portfolio funding objectives. The following paragraphs describe the supply and demand challenges that the portfolio office needs to overcome in order to maximize the throughput of the portfolio in any given funding year.

Supply Challenge #1: Allocating Too Much Too Soon

Consider a situation where a handful of projects (say, ten) were approved for funding during annual planning. Furthermore, consider that each of these projects had a projected demand of $1M for the year. If the portfolio decides to fully fund these projects in the beginning of the year, all the portfolio funds would be used up. No funds would be available if an important new project appears for funding.

Furthermore, as seen in most portfolios, it is quite unlikely that the 10 projects would spend according to their initial estimate (see Demand Challenge #1).

At least some of them will underspend, leading to a surplus. This surplus may come late in the year, making it unusable. Thus, the supply decision of deciding to allocate all the requested demand may result in the portfolio funding fewer impactful efforts and falling short of its optimal output.

Supply Challenge #2: Allocating Funds Too Late

Consider an alternative scenario where the portfolio governance is wary of committing all the funds to early bets like the situation explained in Challenge #1. This wariness might result in a risk averse strategy where the portfolio tends to hold on to the money in anticipation of more projects approaching the portfolio for money. This may result in the portfolio office declining to fund some projects early in the year. However, there is a point beyond which there isn't enough *runway* in the year to execute, even if the projects receive funding. By holding on to the money for too long, the portfolio might hobble itself by not being able to deliver as much impact as it otherwise may have.

Demand Challenge #1:
Underspend in Approved Projects

Projects, even well-managed ones, tend to underspend relative to their budget in any given year. (It's a separate problem that projects tend to spend in excess of their overall budget over the life of the project.) The reasons for underspend include slow ramp-up, delayed availability of key resources, as well as faulty/optimistic estimation of work effort. The end result is that projects hold on to funds far too long into the year and finally realize that they are not going to spend the money. The decision is then made to *turn back* the funds—but this is often done so late that this money is not usable in any practical sense in the portfolio. The portfolio then ends the year with a surplus, resulting in a surprise for the finance department to deal with. It also results in the portfolio achieving less than it could have in that year.

Demand Challenge #2:
Overspend in Approved Projects

While projects in general underspend, most people can recall projects that spectacularly overspent their annual budgets. This can cause problems for the portfolio as outlined; the extra demand for funds from a particular project will mean diversion of funds from other projects and/or reduced funds for new projects. Unless the overspend is caused by valid reasons such as expansion in scope (and consequently expansion in benefits), this creates a less-than-optimal outcome for the portfolio, which is now unable to achieve as much as previously planned.

An additional challenge for the portfolio manager is to communicate the impacts of the previously mentioned challenges to portfolio governance, which is ultimately responsible for making portfolio decisions regarding funding. The portfolio manager needs to advise governance at every step to overcome these pitfalls and enable the portfolio to deliver the maximum possible value to the enterprise.

BEST PRACTICES TO OPTIMIZE PORTFOLIO FUNDING DECISIONS

As we saw in the previous sections, it is almost an art to make the right funding decisions at the right time in order to get the optimal output from the portfolio. However, there are a set of best practices that greatly increase the odds of success in getting the optimal throughput from the portfolio in any given year.

Best Practice #1: Allocation versus Release

Consider a project seeking $100K in funding. In a basic portfolio (or where there is no portfolio), this project would receive the $100K, and that would be the end of it. This effectively removes $100K from the portfolio, and all the challenges mentioned before would apply—namely, the risk remains that the project may not spend the money and may come back at a later time trying to return the money. Now consider a situation where the $100K was earmarked/allocated but only $50K was released. This achieves the following:

- Removes $50K from the portfolio, but keeps track that a total of $100K is needed for this project, and prevents that money from being mistakenly committed to another project, which may lead to an overspend situation for the portfolio
- Preserves $50K for potential deployment for high priority needs that may come up

By this method, the entire sum of $100K is not locked up, and the risk of potential underspend is mitigated by holding on to the $50K for use on other projects. A secondary benefit of this best practice is that it compels the project to come back to portfolio governance for additional funds, during which time the project performance and spend can be reviewed. It's harder to do that once all the funds have been released. Implementing this best practice provides the portfolio with an additional level of control in managing the flow of funds.

Best Practice #2: Managing the "Run-Hot" Factor

What's the *run-hot* factor? Consider a typical portfolio where the projects under-spend their budget in any given year. A portfolio with a $10 million budget that distributes its funds fully to various projects, may see that only $8 million of the money was used at the end of the year. Now imagine that the portfolio, while still only having the same budget of $10 million, approves projects worth $12 million. Again, with the traditional underspend effect in place, the portfolio may see only $10 million of actuals at the end of the year, which is also the exact amount of money that the portfolio had to spend in the first place. Thus, by compensating for the underspend with an extra outlay of money, the portfolio ensured that all funds were deployed to maximum effect. This extra outlay is known as the *run-hot* factor. This is a useful maximizing technique when there is a stable trend of underspend that can be relied upon. Caution—the spend must still be monitored throughout the year to ensure that the spending trend is controlled in line with the *run-hot* assumptions and that the portfolio does not spend beyond its budget.

Best Practice #3: Maintain a Queue of Viable Projects

A portfolio that conducts annual planning will typically produce a list of prior-itized projects (see Chapter 3 for an extensive treatment of annual planning). The projects at the top of the list are funded until the budget limit is reached and the *line* is drawn. The projects below the line constitute the *queue*. These are still viable, useful projects that need additional funds (beyond what was given to the portfolio initially) in order to execute. Now consider funds becoming available during the year due to underspending from other projects. It is very useful to have a ready *queue* of viable proposals that can then be immediately funded to start. Portfolios that do not have a queue will have to launch a hurried search for project ideas and settle for less-than-viable options. What's more, the money will have to stay unused until a suitable project is found.

Best Practice #4: Use the *Turnback* Feature Actively

A project is not inclined to return funds on its own in a timely manner, even if it knows that the funds will not be used. There is a tendency to hold on to the funds, sometimes driven by the mistaken belief that the funds will be carried over into the next year. However, there is a term in portfolio management that denotes the return of funds from the project back to the portfolio—*turnback*. The concept that the allocated funds can be returned may itself be unknown in the organization, and the portfolio manager needs to make everyone aware of this concept and why it is useful (to put unused funds to work while there is still time left in the year to deliver something tangible). The portfolio manager also needs to monitor the spending of projects and actively direct projects with a surplus to turn back funds, which can then be deployed elsewhere. The use of modified earned value management (*mEVM*) (Best Practice #6, described later in this chapter) will greatly enhance the ability of a portfolio manager to spot such underspend trends and initiate feedback requests.

Best Practice #5: Terminate Underperforming Projects

All projects do not perform well, despite the best efforts of the project managers—in fact, there are several respected studies which indicate that a high percentage of projects will underperform. If this is a prevailing trend in projects, why not use it to the portfolio's advantage? The portfolio should double down on winners (well-performing projects) and cut its losses (underperformers). The money saved by terminating underperformers should be reallocated to either well-performing projects with a need for more funds or reallocated to new projects. Project performance management is dealt with at length in Chapter 5. The use of mEVM will greatly enhance the ability of a portfolio manager to spot and track underperformers before they consume an avoidable sum of the portfolio's money.

Best Practice #6: Deploy mEVM

One of the most powerful tools a portfolio manager has in managing a portfolio is mEVM—an objective indicator that measures time and budget performance of a project in terms of concrete deliverables. It has a high degree of predictability of a project's future performance based on its past performance. There is an extensive discussion of mEVM in Chapters 8 through 13 of this book, but in short, for the purpose of managing project funding, mEVM does the following:

- Indicates if a project is performing on time and on budget
- Indicates if a project is holding on to more funds than it needs
- Indicates how much a project will cost at the end with its current spending trend

This information, delivered with precision and confidence, enables the portfolio manager to make correct decisions on portfolio funding and in maximizing the output of the portfolio.

Best Practice #7: Partner with Finance Closely

Finance is the source for all funds in an organization. A successful portfolio will partner with Finance closely and leverage that relationship to its benefit.

Consider a situation where a portfolio takes advantage of a *run-hot* dynamic and funds projects in excess of its budget. Now if the projects spend as planned (none of the projects spend less than planned), and they all perform well (no underperformers that can be terminated), this may actually create a situation where the portfolio runs a risk of spending beyond its budget.

However, if Finance has been taken into confidence and informed well in advance of this scenario happening at the end of the year, they may be able to direct underspend from another part of the organization and meet this potential overspend while still balancing the overall organization's budget. Alternatively, consider a situation where the portfolio will close out the year with an underspend. If the portfolio can inform Finance of this possibility well in advance, Finance can do the previously mentioned mitigation in reverse—take the unspent funds from the portfolio and match it with overspend elsewhere in the organization, again resulting in a net neutral budget outcome for the organization.

Best Practice #8: Monitor OPEX and CAPEX separately

In organizations that differentiate between OPEX and CAPEX (most large organizations treat OPEX and CAPEX separately), there is an added challenge of ensuring that both OPEX and CAPEX usage is optimized and spent according to plan. Consider a project that is underspending its OPEX allocation by $100K and overspending its CAPEX allocation by $100K. Taken together, the two variances cancel out and show that the project is exactly on target and there is no problem here to remediate. However, the reality is that the portfolio office has not one but two different problems to solve.

The solution here consists of matching OPEX and CAPEX variance separately across the portfolio with the hope of canceling out as much as possible within the portfolio. In the event of irreducible variances, the only option left is to go to Best Practice #7, which is to work with Finance by giving them plenty of notice so they can try to mitigate the variance by looking for opportunities in the rest of the organization.

PORTFOLIO FUNDING MODEL

What are the ways in which a portfolio can be funded? There are a couple of models:

- **Centralized portfolio:** The portfolio is given a separate budget from Finance with the expectation that the funds will be disbursed to project proposals in line with the portfolio's objective. What does that mean? Consider a large enterprise with several portfolios—a functional portfolio, infrastructure portfolio, strategic portfolio, etc. The strategic portfolio would accept strategic proposals for funding, whereas an infrastructure portfolio would only focus on proposals seeking to add infrastructure to the enterprise. This model creates a *strong* portfolio that is able to drive compliance with portfolio policies and procedures, because the projects have to approach the portfolio for funds.

- **Distributed portfolio**: In this model, the departments already have the project funds given to them as part of their departmental budgets. The portfolio office only provides nominal oversight and a token approval to use funds that are already with the departments. This model creates a *weak* portfolio—the portfolio has little leverage over departments since they already have the funds. It will be an uphill climb for the portfolio manager to screen proposals and insist on performance criteria for the projects.

LEVELS OF PORTFOLIO MATURITY

Level 1

- No concept of portfolio funding management—there is no awareness that portfolio throughput is a function of portfolio funding
- No concept of turnbacks—money, once allocated, stays with the project till the end of the year whether it is needed or not, creating an inefficiency
- There may not be a practice of annual planning or a queue of proposals waiting for funding
- Portfolio either funds too many projects or holds on to funds until too late in the year
- No active monitoring of underspend or overspend, no attempt to achieve a *precision landing* at the end of the year
- None of the portfolio funding best practices are observed and the portfolio often finishes the year with significant overspend or underspend

Level 2

- Basic awareness of portfolio funding management exists, even if only to avoid exceeding the portfolio budget at the end of the year
- There may only be a limited awareness about maximizing portfolio throughput by applying portfolio funding best practices
- Annual planning is performed, but a well-maintained queue may not be in place to exploit underspend from other projects

- Awareness of the turnback technique exists, although it may be used infrequently
- Portfolio tries to manage funding by being cautious about approving too many projects in the beginning of the year
- There may be some active monitoring of project spend, but no formal way to project year-end cost
- Many of the portfolio best practices may be observed, but the portfolio may finish with a small overspend or underspend despite efforts to avoid variance

Level 3

- Portfolio funding management is actively practiced with a view to maximizing portfolio throughput
- Annual planning is performed effectively and a well-maintained queue is in place to exploit underspend from other projects
- Awareness of the turnback technique exists and is used where appropriate
- Portfolio uses mEVM extensively to forecast project spend, track underperformers, and highlight projects with excess funds
- Underperformers are terminated after appropriate oversight and the funds are returned to the portfolio for redeployment
- Portfolio uses most or all best practices for funding management and finishes with very little or zero variance to the budget

CHAPTER SUMMARY

Management of funds is a central function of the portfolio, and the various nuances around this topic were explored in depth in this chapter. The chapter began with a discussion of the annual funding cycle and the two types of investment funds, OPEX and CAPEX, that each need to be managed and disbursed by the portfolio office. This is followed by a detailed look at the various supply side challenges imposed by the annual funding cycle on the portfolio office's decisions. Next, we explored, in depth, the demand side challenges as well, which also affect the portfolio office's decisions. We then moved on to a comprehensive discussion of the best practices, including the run-hot factor, allocation, release, and turnback features to mitigate both supply side and demand side constraints. Finally, we looked at the different portfolio funding models that are commonly found in organizations and discuss the pros and cons of each model. We ended the chapter with a look at the levels of portfolio capability.

NOTES

1. David Maguire. *The Business Benefits of GIS: an ROI Approach,* 1st ed. Redlands, California: ESRI Press, 2008.
2. Aswath Damodaran. *Applied Corporate Finance: A User's Manual.* John Wiley and Sons, 1999.

5

PORTFOLIO PERFORMANCE MONITORING

INTRODUCTION

Both in financial and project portfolios, there is a prime need to understand how the portfolio is performing. The fundamental reason for performance monitoring is to obtain an early warning of underperformance in order to course-correct to preserve the objectives of the portfolio. In addition to early detection of underperformers, a well-run portfolio also produces a slew of performance-monitoring data that help decision makers make optimal decisions for the portfolio. In this chapter, we cover the following aspects of performance monitoring:

1. A discussion of why projects tend to go off track and how this affects the portfolio
2. A listing of key portfolio indicators that need to be regularly tracked and reported on a periodic basis
3. A discussion of why traditional methods of project performance monitoring are ineffective
4. A basic introduction to the role of modified earned value management (*mEVM*) in project performance monitoring
5. An exploration of the use of the outputs from mEVM for portfolio performance monitoring
6. An introduction to the enhanced monitoring list (EML)

WHAT COULD GO WRONG IN A PROJECT?

The short answer—anything and everything. It is an accepted fact that most projects will run into trouble at some point. The fundamental reason for that lies in the nature of project management: Every project is an experimental endeavor to *create a new product, service, or capability with finite resources and budget within a finite time while meeting concrete performance criteria*. With all these constraints and unknowns in effect, it should come as little surprise that projects get into trouble. When several projects go off course, the portfolio as a whole will start underperforming too. That's why the prudent portfolio office will spot and correct underperforming projects before they lead to an underperforming portfolio. A less advertised function is to terminate projects that are beyond remediation (this will also be covered later in this chapter). Through performance monitoring, a portfolio manager can highlight the following:

- Which projects are performing to plan and which are not?
- How are the off-course projects performing on cost and schedule?
- How is the portfolio performing as a whole—are the underperformers balanced out by a few overperformers?

It's important to note that portfolio monitoring is a value-add activity only when portfolio governance is willing to act on the data.

KEY PORTFOLIO INDICATORS THAT NEED TO BE REPORTED AND TRACKED

Performance data is just one subset of portfolio data—there are other data that the portfolio manager needs to track and make available to governance. Here are the key types of data that governance would be interested in reviewing:

- **Overall portfolio budget**: This data shows how much money has been allocated to the portfolio for the current year. This needs to be broken out by OPEX and CAPEX.
- **Allocated funds versus remaining**: Of the total budget, how much has been allocated to various projects and how much is available for new projects?
- **Allocated versus released**: This data compares the amount of money allocated to various projects versus the amount of money that has actually been released (please refer to Chapter 4 for an explanation of allocation versus release).
- **Actual money spent by the portfolio**: This data shows how much money has actually been spent by the portfolio. This is commonly known as *the*

actuals. Typically, a drilldown should be available so governance can see the actuals by project.

- **Cost performance of the portfolio**: This data shows what the productivity is per dollar spent for the portfolio as a whole (see cost performance indicator in Part II of this book). Here are some additional thoughts on this piece of data:
    1. A drilldown should be available so governance can see the cost performance by project. Sometimes, the cost performance of some well-performing projects could hide or mask the subpar cost performance of some underperforming projects. The drilldown would help to triage which projects are the biggest contributors to the cost variance of the portfolio.
    2. Portfolio cost performance is a useful high-level indicator that can be distributed to executive management to provide confidence that the portfolio is being well run. A trend can also be shown to substantiate how the various portfolio maturity efforts are helping to ramp up the cost performance indicator over time.

- **Schedule performance of the portfolio**: In some situations, timeliness is even more important than cost efficiency. This data shows what the schedule progress is per dollar spent for the portfolio as a whole (see schedule performance indicator in Part II of this book). Here are some additional thoughts on this piece of data:
    1. As seen in the cost performance data, a drilldown should be available so governance can see the schedule performance by project. It may be the case that the schedule performance of some well-performing projects is masking the subpar schedule performance of some underperforming projects. The drilldown would help to triage which projects are the biggest contributors to the schedule variance of the portfolio.
    2. Portfolio schedule performance is another high-level indicator that can be distributed to executive management to provide confidence that the portfolio is being well run. A trend can also be shown to substantiate how the various portfolio maturity efforts are helping to ramp up the schedule performance indicator over time.

- **Estimate to completion (ETC) versus allocated budget**: This data indicator attempts to estimate how much more money is needed by each project requirement before it becomes complete. Here are some additional thoughts on this piece of data:

- The summation of the ETC for all the projects in the portfolio indicates how much additional money is needed by the portfolio for the year. It's very useful to compare this total with the funds still available to the portfolio.
- If the portfolio ETC is larger than the remaining budget, the portfolio is in danger of running out of money. The options are to either trim the spending rate of the projects or to go to Finance and ask for more money.
- If the portfolio ETC is smaller than the remaining budget, the portfolio is in danger of leaving money unspent. The options are to either take up new projects or to inform Finance that the portfolio will be returning some money for potential redeployment elsewhere in the organization.

DRAWBACKS OF TRADITIONAL METHODS OF PROJECT PERFORMANCE MONITORING

As mentioned before, project performance monitoring is key to a portfolio. Most organizations have some kind of project performance monitoring in place. However, these yield very little by way of actionable intelligence in analyzing how the portfolio is actually performing. Here is a look at some of these methods and why a portfolio should evolve past these techniques.

The Notorious R/Y/G System

One of the most popular techniques in reporting project status is the *red, yellow, and green* (R/Y/G) system of reporting, also known as traffic lighting status reporting. While widespread and easy to adopt, this technique adds no value and is prone to misuse because of the ill-defined meanings of R/Y/G. In fact, given the politics of projects, this system is notorious for promoting active misrepresentation of status, as shown by the following anecdotes.

Yellow Is the New Green

Some stakeholders have been found to report their well-performing projects as yellow instead of green. Upon being queried for this unusual request the reasoning usually provided is that they anticipate future performance to drop. The net result is that the status indicator does not represent real performance, but future anticipated performance.

Yellow for Good Intentions

In other instances, stakeholders try to show a *safe* yellow instead of showing red. The reasoning behind that move is generally found to be an optimistic assessment of future performance. Once again, the net result is that the status indicator does not represent real performance, but future anticipated performance. The conclusion of all these ad hoc status selections is that the R/Y/G status colors cannot be trusted to reveal what is really happening with the project.

WHY IS R/Y/G SUCH A POOR PROXY FOR PROJECT PERFORMANCE?

Here are some of the reasons why the R/Y/G system is fundamentally incapable of demonstrating the true status of a project:

- **Reason #1—R/Y/G is mostly subjective**: Consider two projects which are showing *red*. Are they equally underperforming? The odds are that they are not. Going by the color alone leads to a false equivalency, which may result in a suboptimal decision. However, the limitation of the R/Y/G system prevents a further qualification of the color.

- **Reason #2—R/Y/G is vaguely defined**: Consider a project that is performing well on all fronts, except they've completely run out of money—what should they be? For example, the project could be marked yellow because of the underperformance in the budget area, but this may be masking a serious problem in that the project is going to run out of money shortly. In the final analysis, these kinds of questions are hard to answer because there simply isn't enough capacity in the simple R/Y/G system to cover all the different situations.

- **Reason #3—Susceptible to politics**: The simplistic R/Y/G system is easy to manipulate and hence is a favorite for political players within the organization. We saw in the previous section how executives like to change the colors based on the politically convenient message that they want to convey. Amidst all this maneuvering, the true status of the project is lost and this harms the portfolio's ability to make the right decisions.

- **Reason #4—No trending information**: R/Y/G could change every reporting period with no one being the wiser about true project performance. Some organizations try to combat this deficiency by mentioning what the status was in the last reporting period (such as, *previous status: red*). However, this too is of very limited utility and adds little to portfolio governance's quest for visibility about the project's performance history.

In summary, R/Y/G is a very poor indicator of performance and is only to be used as a starting point when there is no other indicator. In other words, R/Y/G is better than nothing, but not by much! Hence the need for mEVM.

mEVM: THE SWISS ARMY KNIFE OF PERFORMANCE MONITORING

Earned value management (EVM) is integral to any discussion of objective performance monitoring. The simplified, modified form of EVM, called mEVM, is elaborated on in Part II of this book. Most of the previously discussed performance monitoring data is sourced from mEVM. The compelling reason to use mEVM is that true performance monitoring needs objective data and mEVM is the easiest way to obtain this in a scalable way. Since mEVM is dealt with extensively in Chapters 8 through 13, we won't go into implementation details of mEVM here; the rest of this chapter will deal with how to use the outputs of mEVM and make the right decisions to improve portfolio performance.

HOW TO USE THE OUTPUTS FROM mEVM FOR PORTFOLIO PERFORMANCE MONITORING

mEVM will indicate one of the following states for each project:

- Project underperforming on cost—performing normally on schedule
- Project underperforming on schedule—performing normally on cost
- Project underperforming on both cost and schedule
- Project performing normally on cost and schedule
- Project overperforming on both cost and schedule

The aforementioned situations are explored in the following section.

#1: Project Underperforming on Cost but Performing Normally on Schedule

Consider a project meeting all its milestones but spending more money than planned. Is this worrisome? It depends on several factors. If the project seems to be getting things done on time, that is always a good thing. Many portfolio managers would not mind paying a little more in exchange for a strategic project that is proceeding on track. However, this could become a problem if several projects in a portfolio start spending more than planned in order to stay on schedule. A prudent portfolio manager would use mEVM outputs to calculate the ETC for every project and find out if the portfolio has funds to support that number (remember to sum up all the ETCs across projects to get to the final number).

Beyond the affordability problem, it's worth investigating why the project is underperforming on cost. It could be a transient problem (the project may get the spending back on track in the next cycle). mEVM will track whether the project has an improving trend or a worsening trend. In the case of a worsening trend, the project is moved to an EML (more details on that later).

#2: Project Underperforming on Schedule but Performing Normally on Cost

Consider a project that has its spending on track but is not meeting its milestones as planned. Is this worrisome? It could be a problem since future spending could increase because the milestones are not met. The total money spent by the project could exceed the project budget. It could also be an issue where the project does not finish per plan and may need to stretch into the next year, taking up funds from the new year's proposals. If several projects in the portfolio have this issue, the portfolio could face trouble down the road.

Again, the solution is to monitor the trend of the projects that are having this issue and put the ones with a worsening trend on the EML.

#3: Project Underperforming on Both Cost and Schedule

Consider a project that is neither meeting its milestones nor able to stay on track with its spending (it's spending more than planned). This is invariably a problem if it persists and is a symptom of serious trouble with either the planning or the execution of the project. Even a few of these projects in a portfolio is cause for concern to the portfolio managers and the governance team. The solution is to put these projects on the EML if the issue persists beyond a monthly cycle.

#4: Project Performing Normally on Cost and Schedule

A project that is delivering its milestones while keeping spending on track is every portfolio manager's dream come true. If most of the projects in the portfolio are in this category, the portfolio is well on its way to becoming a high-performing transformative vehicle. The key is to ensure that projects continue to perform and that any project that is going off course is immediately spotlighted and adjusted to bring it back on track.

#5: Project Overperforming on Both Cost and Schedule

Finally, imagine a project that is performing well on all fronts—meeting all planned milestones on time, but seemingly underspending. While on the surface this looks like a great situation to be in, it actually may not be optimal from a portfolio manager's perspective. The *overachieving* project has more money than it needs, given their rate of spend, and in most cases, projects are reluctant to return money because *they may need it later*.

Imagine the situation that would arise if all projects held on to a *little extra* as insurance. The portfolio will have markedly less funds to allocate to new, viable proposals from the annual planning queue. Therefore, while it's good that the project is meeting milestones on time and under budget, the portfolio manager should direct such projects to return funds as soon as possible so they can be deployed elsewhere.

THE ENHANCED MONITORING LIST

One of the most predictable things in managing a portfolio is that projects will run into trouble and start underperforming. Thanks to mEVM, this underperformance can be spotted early on. However, underperformance can be of two types—the transient *noise* that tends to correct itself and the more stubborn *structural deficiency* that is harder to correct. Before it becomes obvious whether the project's problems are serious or solvable, there is a need to closely monitor the project. In order to facilitate this close monitoring, projects are designated as being on the EML. Typically, this consists of the following:

- An underperforming project is designated as a *monitored* project and is included in the EML of underperforming projects and programs that is reported to governance
- Additional portfolio resources and support provided to all projects on the EML
- Additional performance data on EML projects are reported to the decision makers; this may involve drilldowns to the milestone and deliverable level

Typically, projects on the enhancement list either get better or get terminated. A huge advantage of having the EML is to provide a clear notice to projects to improve their performance or face getting the project terminated. Having such a list in place increases accountability and generally *ups* the game across the portfolio. Additional perspectives regarding the EML are discussed in Chapter 6, which covers portfolio rebalancing.

LEVELS OF PORTFOLIO CAPABILITY

Level 1

- The organization does not understand the importance of project performance monitoring. Consequently, there isn't a standard way of reporting on project performance data. If project reporting is available, it may only be subjective, such as R/Y/G.
- Few portfolio indicators are available, as listed here:
    - The portfolio may not follow an active allocation strategy. Therefore, it may be hard to estimate how much money has been allocated to projects and how much is left to be allocated.
    - The portfolio may not employ a strategy of allocation versus release and may simply release whatever is allocated. Therefore, it may not be possible to obtain allocation versus release as an indicator.
    - Since project actuals may not be available, there may not be a way of showing the actual money spent by the portfolio as a whole.
    - Since mEVM is not implemented, (and furthermore, project actuals are not available), there is no way to show cost performance or schedule performance of the portfolio or of individual projects.
    - The ETC may not be tracked for each project and hence, is not available for the portfolio either. Consequently, it may not be possible to show ETC versus allocated budget for the portfolio.
- The only project monitoring system in place may be the R/Y/G system, which could mean that there is no real information about project progress or milestone status. There is also no info about whether a project is overspending or underspending
- The concept of an EML is unknown or the organization has not decided to implement it.

Level 2

- The organization understands the importance of project performance monitoring. Consequently, there is a standard way of reporting on project performance data. Although this may still be a subjective measure such as R/Y/G, it is applied in a uniform manner with many of the typical R/Y/G drawbacks mitigated.
- Several meaningful portfolio indicators are available and are published as listed here:

- The portfolio office employs an active allocation strategy. Therefore, it is possible to estimate how much money has been allocated to projects and how much is left to be allocated.
- The portfolio office employs a strategy of allocation versus release and does not release the entire allocated amount at once. Therefore, it is possible to obtain allocation versus release for each project as well as the overall portfolio.
- Project actuals are available and compared to the approved budget. Some kind of spend plan exists and actual spend versus planned spend can be tracked. This also means that the actual money spent by the portfolio as a whole can be shown.
- Since mEVM may not yet be implemented, there is no way to show cost performance or schedule performance of the portfolio or of individual projects.
- The ETC may be tracked for each project, but since it is not backed up by an objective indicator such as mEVM, the confidence in the ETC might be low.
- The ETCs for the projects could be summed up to show an ETC number for the portfolio. Consequently, it may be possible to show ETC versus allocated budget for the portfolio.

- The only project monitoring system in place may be the R/Y/G system, which could mean that there is no objective information about project progress or milestone status. There is also little confidence about whether the project is overspending or underspending.
- Although the concept of the EML is known to the organization, it may not be possible to implement it without an objective indicator such as mEVM being in place.
- Although some meaningful portfolio data is available as previously shown, action may not be consistently taken in response to the data.
- There may be a comprehensive report on portfolio data and/or portfolio performance; however, it may not have look-ahead capability.

Level 3

- The organization fully understands the importance of project performance monitoring. Consequently, there is a standard, objective way of reporting on project performance data, such as having mEVM in place. The presence of an objective measurement indicator removes a lot of the limitations that are seen in Levels 1 and 2.
- A full host of portfolio indicators are available, and are published as listed here:

- The portfolio office employs an active allocation strategy. Therefore, it is possible to estimate how much money has been allocated to projects and how much is left to be allocated.
- The portfolio office employs a strategy of allocation versus release and does not release the entire allocated amount at once. Therefore, it is possible to obtain allocation versus release for each project as well as the overall portfolio.
- Project actuals are available and compared to the approved budget. An mEVM-backed spend plan exists and cost variances can be tracked. This also means that the actual money spent by the portfolio as a whole can be shown.
- Since mEVM is fully implemented, both cost performance and schedule performance of each project can be displayed. These measures can be aggregated to show the cost and schedule performance of the whole portfolio.
- The ETC is tracked for each project and is backed up by an objective indicator, such as mEVM. This creates a high confidence in the ETC number.
- The ETCs for the projects could be summed up to show an ETC number for the portfolio. Consequently, it is possible to show ETC versus allocated budget for the portfolio.

- Since mEVM is the project monitoring system in place, objective information is available about project progress or milestone status. There are clear indications about whether the project is overspending or underspending.
- An active EML backed by mEVM exists and serves as an effective measure to manage troubled projects.
- There is a governance body that has the political will to act on the strength of the aforementioned objective data.

CHAPTER SUMMARY

In this chapter, we explored the essential components of portfolio data that need to be monitored. We also addressed the core issues of monitoring project performance along with an analysis of the shortcomings of traditional reporting methods. We continued with an introduction of mEVM, a powerful technique that is the cornerstone of objective performance reporting. It concluded by covering the topic of enhanced monitoring and the need for portfolio governance to act on the gathered performance data.

6

PORTFOLIO REBALANCING

INTRODUCTION

The concept of portfolio rebalancing has its origins in financial portfolio theory, where rebalancing is the process of realigning the holdings of the portfolio. In this chapter, we will cover the following:

1. Approaching portfolio rebalancing by looking at the concept in financial portfolios
2. Porting the concept of portfolio rebalancing to project portfolios
3. Defining the fundamental need for project portfolio rebalancing
4. Describing portfolio scenarios where rebalancing is appropriate
5. Exploring why portfolios find it hard to perform rebalancing
6. Considering an alternate view of rebalancing
7. Looking at portfolio levels of maturity

PORTFOLIO BALANCING IN A FINANCIAL PORTFOLIO

The easiest way to approach the concept of portfolio rebalancing is by first illustrating how it works in a financial portfolio, since most people are acquainted with one. In a financial portfolio, portfolio rebalancing typically involves divesting some assets and buying other assets in order to maintain an original desired goal or benchmark, as in a retirement portfolio whose target asset allocation is a 1:1 mix between stocks and bonds. If the stock holdings grew as a result of solid growth during the period, it may wind up becoming 75% of the portfolio. The asset manager then sells some stocks and buys some of the bonds to get the portfolio back to the original target allocation of 50/50.

APPLICABILITY OF PORTFOLIO BALANCING TO A PROJECT PORTFOLIO

Does portfolio rebalancing apply to a project portfolio? Definitely, but in a markedly different way from how it applies to a financial portfolio.

A project portfolio contains projects—not financial assets such as stocks and bonds in a financial portfolio. So, the concept of growth does not apply here—a project doesn't *grow* like a stock or bond over time. However, like a financial asset, a project can be considered to be a *performing* or a *nonperforming* asset. Here are some ways that a project can be a *nonperforming* asset:

- Project performance is consistently below par—not meeting cost or schedule expectations
- The project is not delivering promised benefits
- The project is no longer a strategic fit—the strategic landscape changed after the project was approved

In this aspect, a project portfolio needs to adopt the mindset of an actively managed fund that prizes returns above all else. In a financial portfolio, all that matters is performance/returns. Losing positions (nonperforming investments) are cut with alacrity and the funds are redeployed into other positions—new or existing. While the subject of *project portfolio versus financial portfolio* has been extensively debated, in this particular aspect, project portfolios would do well to emulate the financial portfolio model—cut the losses and let the winners run.

THE FUNDAMENTAL NEED FOR PORTFOLIO REBALANCING

The fundamental need for portfolio rebalancing comes from a need to react to change. This change could come in many ways—a project could turn out differently than planned or the strategic direction of the company could necessitate a pivot toward new projects. Other drivers of change could be a belt-tightening exercise, where the portfolio may be directed to execute a 10% budget cut or a directive to self-fund a new critical priority. Change may not always be negative—there could be availability of new surplus funds, or certain projects may find that they don't need as much money to accomplish their objectives. Whatever the driver of change may be, a portfolio should always be in a position to exploit the same to its advantage.

THE PREREQUISITES OF PORTFOLIO REBALANCING

Portfolio rebalancing is hard to perform unless the following prerequisites are in place:

1. **Enhanced monitoring list (EML)**: A special list of underperforming projects and programs that receives additional scrutiny from governance.
2. **A project queue**: A *queue* is a collection of vetted projects with positive return on investment (ROI) and strategic fit that fell below the *affordability line* during annual planning.
3. **Modified earned value management (*mEVM*) implementation**: An objective system of measuring project cost and schedule performance, which is described extensively in Part II of this book.

SCENARIOS WHERE PORTFOLIO REBALANCING IS REQUIRED

In the following sections the most common portfolio rebalancing scenarios are explored.

Scenario 1: The Underperforming Project

As explained in previous chapters, the one predictable situation that a portfolio manager can expect to encounter is that of dealing with a project that is not performing to cost and schedule expectations. While some projects work through their challenges and come back to perform as planned, others are not able to do the same and need to be put on an EML.

Regrettably, there are bound to be projects that are not able to recover in spite of being on the EML. The only option left in that case is to terminate the project and redirect the remaining funds. What is the best way to decide how to redirect the funds? Here are some options:

- Recall from previous chapters the need for maintaining a queue of validated projects that are *solid bets* for funding. This would be an ideal use of the funds that are freed up by terminating underperforming projects.
- Some existing projects may be in need of additional funds. You may recall that in Chapter 4 the recommendation was to *not* release the entire amount of funds requested by a project all at one time. Such projects (ones that only had a partial release of the total funds requested) may now be asking for additional funds. Provided that these projects are in

good standing from a cost and schedule perspective, these become viable recipients of the funds made available by terminating underperforming projects with a deteriorating trend.

Scenario 2: The Famine

Another recurring situation that a portfolio manager has to contend with is the belt-tightening exercise. Due to the cyclical nature of the business cycle, portfolios are sometimes asked to *give back* a percentage of their budget to help cover a shortfall elsewhere in the business. Typically, portfolios deal with such requests by resorting to *an across-the-board cut*—all projects are asked to share the pain by taking a percentage cut. While this seems democratic and fair, in reality this is quite suboptimal for the portfolio because not all the projects are performing at the same level. There are some projects that could be completely terminated without any impact and others that might have a severe impact to the strategic journey if their funding was reduced. Here are some options on how to approach this situation:

- Underperforming projects on the EML are good candidates for termination and recovery of funds, especially if the project's cost and schedule variance trend is worsening. If these projects are eventually going to get terminated, this might be a good opportunity to go with that decision.
- Solidly performing projects that seem to be underspending (see scenario of project underspending in Chapter 4) could give back funds without affecting their progress.
- If additional cuts are needed even after one of the previous two options was implemented, consider spreading the cuts across less strategically important projects and preserve the current level of funding for the truly strategically critical projects.

Scenario 3: The Feast

A somewhat less frequent scenario to contend with for the portfolio manager is the sudden availability of additional funds toward the latter half of the year. A well-run portfolio will not normally have unused funds piling up as the year progresses because of the emphasis on optimal fund usage throughout the year. However, other portfolios within the organization might not have this rigor and might find themselves with more money than can be deployed. Alternatively, unspent funds from the rest of the organization might be *looking for a home*. The advantage of having a well-run portfolio is that it is ready to take advantage of such funds made available by inefficiencies elsewhere. Here are some options to approach this situation:

- The importance of having a *queue* makes all the difference here: To refresh your memory, a *queue* is a collection of vetted projects with a positive ROI and a strategic fit that fell below the *affordability line* during annual planning. With the availability of additional funds, these are prime candidates to receive funding. *Please note* that it may be necessary to prioritize the queue to pick *short life, quick win* projects. The reason is that any excess funding is available for that financial year alone and should not be counted to continue into the following year. Thus, it is best at this point to pick projects from the queue that can get done quickly.
- Some projects can be accelerated: mEVM readouts would indicate projects that could execute faster with more funding. (Overachieve on schedule variance at the cost of reduced performance on cost variance.) Such projects would be great candidates for additional funding because the demand for funds would be reduced in the new year (since more of the project completion is being accelerated this year).
- Investment in portfolio tools and capability enhancement: A well-run portfolio will always have plans to get to the next level in each capability journey. The availability of excess funds would be a good opportunity to execute on the plans and create a force-multiplier effect for the portfolio.

Scenario 4: The Strategic Reset

Sometimes the organization will pivot in its strategic positioning. In a well-run portfolio this will have a direct impact on what projects are being worked on— because of the close, direct link between the projects and the strategy being pursued. Therefore, the portfolio manager needs to be ready to take actions to reorient the portfolio to align with the new strategy. Here are some options for consideration:

- The strategic road map will need to recast and the existing projects will need to be reranked in terms of priority—with the help of the priority advocates (see Chapter 3).
- The low-ranked project under the new strategic direction will need to be treated as follows: Try to complete if near to the end and if not possible, plan to terminate so as to conserve funds and bandwidth for the execution of new projects ranked high in the strategic setup.
- In addition to the previous considerations, projects that were underperforming in the old strategic direction should also be candidates for termination to allow for redirection of funds and manpower to the new projects.

In all of the scenarios mentioned here, the common requirements are the need for mEVM reporting and the need for robust annual planning along with the constituent deliverables such as the strategic road map. Having these in place makes it straightforward to perform the course corrections needed as part of portfolio rebalancing.

WHY PORTFOLIOS FIND IT HARD TO REBALANCE

Portfolio rebalancing is a great technique to ensure that the portfolio is always performing at its peak potential in delivering strategic change. It is a powerful tool that can be used to deal with unexpected change. However, most portfolios find it hard to execute this maneuver. Here we explore some of the reasons why.

Reason 1: Lack of a Venture Capital Mindset

What's a venture capital mindset? Quite simply, a venture capital funds mindset is to bet on many startups, hoping that some of them will pay off significantly. While all care is taken to vet the startup prior to funding, it still remains a speculative activity. Many startups fail and get terminated—this is the norm and all players are aware of it. This consequently creates a *win-or-die* dynamic where the winning startups receive more funds at the expense of concepts that were rejected by the market. While it's true that a project portfolio has many constraints that are not typically faced by a venture capital fund, this *win or fail quickly* is something a portfolio would profit greatly from by adopting. Projects are selected and planned with due diligence, but underperformers need to be quickly terminated so that resources can be focused on new projects and existing successful ones. *Nonperforming projects are at risk of termination* is the mindset that the rank and file need to adopt.

Reason 2: Reluctance to Kill Projects

Organizations are reluctant to kill projects, even the obvious disasters. This *hesitate-to-terminate* mindset is one of the single biggest obstacles that could keep a portfolio from reaching its potential. Why? Quite simply, a portfolio has finite resources—both financial and manpower. Unless underperforming projects are terminated, there is no way to start new, promising projects. The only other way to start new projects is to wait for current projects to conclude on their own, which may take a long time, especially if there is no pressure to perform. It's a vicious circle that reinforces mediocrity and prevents a portfolio from rising to its full potential.

Reason 3: Lack of Data to Make the Right Decisions in Rebalancing

One primary reason why underperforming projects survive so long is that there is no objective measurement of performance. Using traditional ineffectual systems—such as the red, yellow, and green lighting—yields almost no value in spotting underperformers. Even after a project is visible to all as an underperformer, a lot of time is wasted in *waiting for the project to turn around*. In other words, no one knows if a turnaround is working. Given all the aforementioned reasons, rebalancing in the absence of objective data is very challenging.

Reason 4: There Isn't a Queue of Viable Projects

Critical to the rebalancing concept is the availability of a *queue*—a list of viable replacement projects which are *good bets*. However, few organizations have this queue ready—partly because having such a queue necessitates having a robust annual planning process, which is another capability that few organizations have. By now, it should be apparent how interlinked portfolio capability dimensions are—getting to the next level calls for many prerequisites.

Reason 5: There Isn't an *Enhanced Monitoring List* or *Kill List*

An EML list or kill list is a great source for rebalancing opportunities. Why? Because the projects on that list have landed there for a reason and, in some cases, are unable to get back on track in spite of enhanced support and review. However, few organizations have an EML list or kill list. The reason again is the lack of objective data to designate projects as candidates for the EML list, which leads to a self-reinforcing cycle.

Reason 6: Rebalancing Is Political and Hence Avoided as Far as Possible

In the absence of (a) a performance-based mindset, (b) a tradition of regular rebalancing, and (c) objective data to make termination decisions, portfolio rebalancing becomes a political minefield. Stakeholders of underperforming projects are likely to push back hard at any moves to terminate their projects. Few leaders want to waste their political capital fighting these battles, which seem arbitrary in the absence of well-established criteria.

AN ALTERNATIVE VIEW ABOUT PROJECT PORTFOLIO REBALANCING

Based on all the preceding discussion, it would seem that portfolio rebalancing is primarily concerned with *returns*—projects that deliver results continue to be funded and projects that don't are terminated to make way for more projects that promise to deliver. While this is how most portfolios want to operate (we need to do worthwhile projects—why do projects that may waste the organization's money?), sometimes there are other considerations as to why we would take a different approach.

Consider a *diversified* financial portfolio—it is invested in a variety of instruments aimed at reducing risk/exposure while delivering promised returns to the investor. In contrast, a diversified project portfolio is aimed at meeting multiple objectives for the organization and correspondingly consists of different kinds of projects. If some of these projects have to be terminated as part of rebalancing, the portfolio is still constrained to choose other projects of the same type it just terminated—it may not be at liberty to choose the next-best proposal (financially).

To illustrate, consider a diversified portfolio whose composition has 60% grow-the-business (GTB) projects and 40% run-the-business (RTB) projects (imagine the board of directors has tasked the company leadership to maintain this mix of *grow* and *run*). Suppose several GTB projects are found to be underperforming and need to be terminated. The portfolio may have a steady queue of promising RTB projects waiting for funding. Can the portfolio then pick up a bunch of RTB projects in addition to the existing projects and turn the portfolio composition mix to 90% RTB and 10% GTB? That may not be feasible from an organizational point of view. So the organization may be constrained to pass over the promising RTB projects and fund the next GTB project in the queue as part of rebalancing.

Alternatively, governance may approve—for now—the funding of several RTB projects with compelling value propositions but may direct that the portfolio come back to its desired composition mix at the next balancing opportunity. There's no one correct way—each organization will have to consider the factors important to them and make decisions correspondingly.

Levels of Portfolio Capability Maturity

Level 1

- The organization is unfamiliar with the concept of portfolio rebalancing—it's unusual to terminate a project once it gets started

- Organizational culture does not support rebalancing—projects just keep moving along regardless of the relative importance to the portfolio's objective
- New projects are only able to begin with new funding or when existing projects complete
- There isn't an objective assessment of project performance to support any rebalancing decisions
- None of the prerequisites for rebalancing are in place, which makes any rebalancing activity sporadic and nonuniform

Level 2

- An awareness of project rebalancing exists but it is not a well-grounded concept—hence, it is not done uniformly or comprehensively
- Rebalancing may only be done at the end of the year and is usually driven by funding availability
- If done during the year, rebalancing typically takes the form of *across-the-board* cuts that were necessitated by funding shortfalls
- Rebalancing is unlikely to be driven by objective measures of performance (such as mEVM)
- Other prerequisites for productive rebalancing such as mEVM, EML, and a project queue may not be present

Level 3

- Portfolio balancing is a well-known and embraced concept—rebalancing is done at regular intervals
- There is awareness and expectation of regular portfolio rebalancing, and this is a driver for accountability and project success
- To support rebalancing, all of the prerequisites exist—namely mEVM, the EML, and a project queue
- Rebalancing also is taken into account during one-time events such as portfolio funding reductions as well as distribution of surplus funds

CHAPTER SUMMARY

In this chapter, we explored the vital activity of portfolio rebalancing, which helps portfolios cope with change. We approached the concept by looking at how portfolio rebalancing is done in financial portfolios. We then ported the concept of portfolio rebalancing to project portfolios and proceeded to describe the fundamental need for project portfolio rebalancing. We further explored portfolio scenarios where rebalancing is appropriate while also considering the real-life constraints that make it hard for portfolios to perform rebalancing. We also considered an alternate view of rebalancing before rounding out the chapter with a look at portfolio levels of maturity.

7

PORTFOLIO BENEFITS REALIZATION

INTRODUCTION

What is the fundamental reason to execute a project? Quite simply, organizations execute projects for the benefits delivered by the projects. A benefit is an outcome or end result that is advantageous to the organization. It could be increased revenue, increased market share, customer satisfaction, decreased cost, and so on. Therefore a high-performing portfolio is created by selecting projects that each deliver tangible benefits to the organization.

However, that is harder to do than it appears—in reality, most projects that are funded simply do not have solid benefits that stand up to scrutiny. Implementing a benefits management process is a hugely effective filter to screen out projects that don't deliver benefits. In this chapter, we cover the following aspects of benefits realization:

1. Classifying the different types of project benefits
2. Describing the need for a portfolio benefits management process
3. Describing the building blocks of the portfolio benefits management process
4. Describing the obstacles to benefits attainment and how to mitigate these obstacles

TYPES OF BENEFITS

Monetary Benefits

- **Hard return on investment (ROI):** is typically either an actual reduction in costs or an actual increase in revenue. Hard ROI needs to be recognized as such by Finance and should be directly attributable to the project. This kind of ROI is most prized by portfolio managers, and any project with a positive, verified ROI is a strong candidate for funding. *Please note* that it's also rare to find projects with credible hard ROI benefits.

- **Cost avoidance:**[1] refers to reductions that cause future spending to fall, but not below the level of current spending. Often, cost avoidance involves slowing the rate of cost increases. In other words, future spending would have increased *even more* in the absence of cost avoidance measures. While sometimes attractive, projects whose ROI is cost avoidance need to be scrutinized carefully. A project that claims to avoid a future increase must be able to demonstrate a rising trend in costs that is then flattened due to the execution of this project. *Please note* that cost avoidance projects tend to have optimistic projections that may be hard to substantiate.
- **Soft savings:** are typically created by measuring an increase in productivity or an increase in the efficiency of an employee or process. It's not exactly a monetary benefit, but the increase in productivity can (and often is) translated into a dollar figure by converting hours saved into dollars. Soft savings usually are the bane of portfolio managers everywhere, because most projects that are not worth doing often have benefits purporting to be soft savings.

NONMONETARY BENEFITS

Sometimes, the benefit of doing a project may be strategic or delivering a capability to the organization. It may also be aligned with delivering objectives that are important to the business. While some of these nonmonetary benefits can be quantified in monetary terms, at times the project may have to proceed on the basis of the nonmonetary benefits. In that case, it's important to ensure that the nonmonetary benefits are still worthwhile.

It takes some planning and screening to ensure that the project's benefits, while nonmonetary, are indeed aligned with the strategic road map of the organization. Another key factor in dealing with nonmonetary benefits is to ensure that business signoff exists before starting the project.

THE PORTFOLIO BENEFITS REALIZATION PROCESS

Need for a Portfolio Benefits Realization Process

- **Need #1:** People need to have a standard understanding of benefits. Often, people do not understand the project benefits—every project is *beneficial/worth doing* in the eyes of the respective project owner.

- **Need #2:** Periodic project reviews are essential. Without a periodic review, the project benefits realized (if they are realized at all) have no relation to the benefits promised during project selection. Projects have a tendency to exaggerate benefits during the proposal stage.
- **Need #3:** There is a need for a mechanism to hold projects accountable for promised benefits. Projects are never going to self-regulate in terms of benefits delivery.
- **Need #4:** The existence of a standard and periodically scheduled process to review all projects' benefits creates a climate of transparency and responsibility on the part of projects to deliver on their promises. This also creates a mechanism to spot projects that are unable to deliver.

Building Blocks of a Portfolio Benefits Realization Process

Step #1: Uniform Taxonomy to Classify Benefits

The foundation for a successful benefits realization process is for all of the stakeholders to speak the same language when it comes to benefits. This, is turn, makes it necessary for a document to describe the standard types of benefits that are used for that organization. Consider Table 7.1—this table is a small sample of what the benefit taxonomy chart (BTC) would look like. Every benefit category is derived from a strategic priority and further classified as monetary or nonmonetary. (Monetary entries could be further divided into hard and soft.) A description column is provided to capture additional nuances. The table shown here needs to be completed and socialized with the business partners as well as other functions such as Finance (especially the monetary information). Once everyone can agree that this document captures all of the relevant benefit types, it serves as the BTC. All proposals that promise benefits will then need those benefits to be classified according to this chart.

Table 7.1 Basic form of a BTC

Strategic Priority	Benefit Category	Monetary/ Nonmonetary	Description/Notes
Profit growth	Increase in sales	Monetary	
	Increase in margin	Monetary	
Customer service	High availability	Nonmonetary	
	Customer satisfaction	Nonmonetary	

Step #2: Mandatory Declaration of Project Benefits in Standard Form

Surprisingly, in most organizations, benefits are somewhat of an afterthought when project proposals are floated. In some cases, projects try to avoid talking about benefits by promising to provide benefit information afterward. In other cases, projects create ad hoc, vaguely worded benefits that are not easy to validate or verify after go-live. The remedy for those situations is to make it mandatory for project proposals to choose benefits from the BTC. In other words, a project can only pick something that is already on the BTC—the project is not at liberty to make up its own benefit.

The choice of benefits can happen in two possible ways—during the annual planning process, the template that is filled out needs to contain choices from the BTC. For off-cycle proposals, the intake process needs to be integrated with the BTC such that the proposals contain the benefit information.

Step #3: Identification of the Earliest Time That Benefits Can Be Reviewed for Each Project

Projects have a tendency to postpone benefit review until after go-live. In other words, they do not want to engage in any discussion about whether the promised benefits are materializing until it's a *done deal* (which is project go-live). At that time, it's mostly an academic exercise, because the money is already spent and there is nothing that can be done, even if the project benefits are nonexistent.

The approach of reviewing benefits after project go-live is contrary to the interests of the portfolio. A portfolio manager should seek to terminate projects that are not likely to deliver any benefits and redirect those funds to existing projects with tangible benefits or to new projects with promising benefits. To enable this to happen, the earliest time one or more benefits belonging to the project can be reviewed should be identified (along with project owner commitment to the same).

Step #4: Securing Agreement about Project Benefits with the Project Sponsor

Project sponsors (typically senior people) will often find their names applied to various proposals and may not really stand behind that proposal—simply because no one bothered to share the details of the projects with them. This is especially true in the case of project benefits; rather generous promises are made regarding benefits and the implicit understanding is that the project sponsor supports those claims.

When the project ends, and the benefits are typically less than promised, it creates an awkward feeling all-around. Some organizations deal with this problem in the following way—once the benefit projection is created for a project and some kind of validation/backup is established to support these projections, an agreement is created with the project sponsor listing the benefits and requiring the sponsor to *sign off* on the same.

It's also typical for these agreement documents to have supporting statements from Finance and other business partners regarding the benefits. This is a good checkpoint for the sponsors to review and agree to the documents that purport to bear their name while promising benefits to the rest of the organization. It adds credibility to the benefits promised and is an indication of due diligence having been performed.

Step #5: A Functioning Portfolio Benefits Review Council

A portfolio benefits review council is a governance body that focuses on reviewing the promised benefits of a project. It could be a subcommittee of the larger portfolio governance body. The functions of a benefits review council are:

- Periodically review the benefit information of each project and compare to the promised level of benefit at the time of project approval
- In case of significant variance (council to decide how much is significant), decide on whether to place this project on the enhanced monitoring list (EML)
- For the projects that are on the EML due to an under delivery of benefits, continue to monitor benefits attainment and make a recommendation for exit from the EML (return to normal status or termination of project)

The portfolio benefits review council is either part of the main portfolio governance council or is a subcommittee that reports to the main portfolio council. Some key pointers to the functioning of the portfolio benefits review council are:

- Having this council in place helps greatly with tracking benefits for all projects. Now that there is a dedicated body that reviews benefits attainment (at least quarterly), the projects are compelled to treat this topic seriously—as opposed to the historical pattern of either ignoring benefits or making them up.
- For this council to be most effective, project benefits need to be reviewed well before projects go-live (see Step #3 the in previous text).
- Some organizations mandate that only projects that are bigger than a defined threshold should go to portfolio benefits review. This is done to triage the projects to ensure focus on the *big-spend, big-impact items.*

Step #6: Regular Report-Out on the Benefits

If all of the preceding steps are followed, the organization can reasonably expect to see a steady stream of validated benefits. The next logical step is to report out on these benefits achieved by the portfolio. Some of the numerous benefits include:

- Demonstrating the value of the portfolio to the business and other partner functions. This creates value substantiation during annual budget planning activities.
- Creating a culture where benefits are expected and celebrated—as opposed to being an optional afterthought.
- Creating accountability for projects that are promising benefits. It ensures that the projects will need to stand behind their original benefit projections.

Obstacles to Benefits Attainment

This section is about the typical obstacles in the way of portfolio managers trying to implement a robust benefits attainment program for the portfolio.

Obstacle #1: Projects Avoid Mention of Any Benefits
While Seeking Funding

One of the ways people try to defeat benefits realization is to not even provide the project benefits while requesting funding. People claim that they don't have enough data to provide benefits until they receive funding to do the project. While there is some truth to the fact that it takes work to perform analysis on the impact and benefits of a project, it's a classic red flag for a project to request major funding or full funding without providing benefits.

The prudent way of handling the situation is to provide a project with a standard minimal allocation—say 500 man-hours (this is a middle of the road estimate, and would need to be adjusted as needed)—to provide a comprehensive benefits statement. Second, funding must be contingent on the project integrating itself with the benefits realization process. This means that projects that do not appear for benefits review would have further funding withheld.

Obstacle #2: Overstated Benefits to Make the Proposal More Attractive

Other projects promise vastly optimistic benefits that have no relation to actual ground reality. This is done to create a compelling value statement to increase funding probability. The remedy for this situation would be to hold a formal benefits realization review—as described in Steps 3, 4, and 5 of the previous section. At this review, a comprehensive look is taken at the actual benefits being delivered by the project as compared to the projections made at the time of funding.

Periodic benefit reviews need to start well before the end of the project. A warning flag for underperforming projects is an inability to show any benefits before the completion of the project. Such projects need to be revisited and the question of viability reconsidered.

Obstacle #3: Allowing Project Performance to Be a Proxy for Benefit Delivery

Some projects put forth their on-time, on-schedule performance as a proxy for benefits attainment. The point of view that's being advanced is along the lines of: *the project is doing well, so you can expect the benefits to be delivered as promised, too.* This is a fallacy because the project being well run does not mean the outcome of the project will deliver the benefits. The project could execute on-time and on-budget and the end product could still deliver none of the promised benefits. The solution is still as outlined before—insist on early identification of benefits and also regular monitoring of benefits through the portfolio benefits review council.

LEVELS OF MATURITY

Level 1

- No concept of benefits management
- Project benefits are ad hoc—there is no standard framework to describe benefits
- There is a wide disparity in project benefits—some projects have no benefits and are still approved
- All benefits are lumped together—there is no distinction between hard and soft benefits, or between cost avoidance and cost savings
- Projects that actually have a potential to deliver hard benefits and actual cost savings are not favorably funded over others that have more modest benefits
- Some projects have vastly optimistic benefits that are never held to account
- No standard benefits review process—any review is done in brief, as part of the project postmortem

Level 2

- An awareness of the need for project benefits exists but is not a well-grounded concept—hence, it is not done uniformly or comprehensively
- Some kind of standardization exists with regard to project benefits but there are many exceptions to the rule
- No partnership with Finance and/or the business regarding validation of business benefits
- Some disparity exists regarding project benefits—a few projects are still approved with slim or nonexistent benefits
- Some projects have vastly overstated benefits that are never held to account
- There may be a standard benefits monitoring process but it usually happens after go-live and is ineffective in stopping projects that don't deliver benefits

Level 3

- A strong awareness exists about the need for project benefits and it is well integrated into the whole project/portfolio life cycle
- There is a standard benefits taxonomy and all projects are made to adhere to it
- Projects that show potential hard benefits and actual cost savings are prioritized over other projects with more modest benefit profiles
- A robust partnership with Finance and the business regarding validation of business benefits
- All projects follow the benefits attainment process—no projects are allowed to proceed without a validated benefits statement reviewed by governance
- There is a standard, robust benefits monitoring process that monitors project benefits before go-live and is effective in stopping projects that won't deliver benefits, and projects with overstated benefits are held accountable to delivery of those benefits

CHAPTER SUMMARY

In this chapter, we introduced the important topic of project benefits. First, we explored the two major types of benefits—monetary and nonmonetary. We proceeded to outline the structure of a portfolio benefits realization process and walked through the building blocks of the process. We then covered the typical obstacles in the way of portfolio benefits realization and how to handle these obstacles. We finished with identifying the capabilities at each level of portfolio maturity.

NOTE

1. *The Cost Benefit Knowledge Bank,* http://cbkb.org

Part II

mEVM

8

INTRODUCTION TO mEVM

INTRODUCTION

The biggest problem in portfolio management is the difficulty in making data-based decisions. By way of introduction, let's compare this situation to a financial portfolio. While we've made many comparisons between project portfolios and financial portfolios, financial portfolios have one impressive advantage over project portfolios—the performance of the financial portfolio is transparently clear and updated daily at every market close. There is never room for doubt as to whether the portfolio has made money or lost it. Furthermore, it's apparent which holdings are making money and which are bleeding cash. This then enables a very important decision—which investments to close out and which ones to invest more in.

While this visibility is ubiquitous in finance portfolios, it is quite rare in project portfolios—not only is it hard to know how the whole portfolio is performing, it is also quite hard to know which projects are performing well and which are not. We're talking about objectively measuring performance—many portfolios have pseudo measures which may not be accurate (see Chapter 5). Because of this prevalent dynamic in project portfolios, it's difficult to make data-based decisions. It's hard to *cut your losses and let your winners run* because it's so hard to tell which project, really, is a winner or a loser. This chapter covers the following aspects of performance measurement:

1. Explores the fundamental difficulty in monitoring portfolio performance
2. Describes the inadequacy of traditional project reporting
3. Introduces modified earned value management (*mEVM*), a powerful technique that allows portfolios to make data-based decisions

4. Describes the origins of earned value management and proceeds to describe the simplified variant of mEVM

WHY IS IT SO HARD TO REALLY SEE HOW THE PORTFOLIO IS ACTUALLY PERFORMING?

Some of the most common reasons are described here:

- **Reason #1—Active misrepresentation**: As covered in Chapter 5, a high percentage of projects are in trouble at any given time. This is not surprising—it's inherent to the nature of the undertaking. But no project wants to broadcast their *not doing well* status to the world. Therefore, while they attempt to fix their project, the safe status seems to be *doing okay for now*.
- **Reason #2—Natural chaos**: Often, the project leadership themselves may be unaware of their true status. If the team members report an artificially healthy outlook, the project manager packages this up and sends it to the portfolio. In the absence of an objective indicator, no one really knows how they are doing.
- **Reason #3—Ineffective systems and processes**: In many cases, the available systems and processes may simply be inadequate for reporting the true status. The inadequacy of the red, yellow, and green system is an example. Another example is qualitative reporting of status, that is, a description of status, in words, that change from period to period to allow for zero retention and follow up. Often, the governance body is simply unable to understand, process, and take action based on the quality of status reporting. Not having a trend or numbers to work with creates a hobbling effect.
- **Reason #4—Politics**: Portfolio management can be a politically charged topic, as it involves budget and resource allocation. Consequently, portfolio status reporting becomes quite political, too. Everyone wants to keep parity with the other political players in controlling the message about their project's performance. Few people want to be the outlier who reports the true status while everyone else projects an artificially healthy picture. Poor visibility *helps* everyone except the portfolio manager and executive management.
- **Reason #5—Politeness**: Given how the corporate world works, the established practice is to accept people at their word until data to the contrary is too strong to ignore. But by the time accumulated contrarian data is obtainable, the damage may already be done. This is especially true when it comes to project performance.

TRADITIONAL REPORTING SOLUTIONS AND THEIR DISADVANTAGES

Some organizations are aware that all of the previously mentioned factors may inhibit true status reporting. The traditional solution, tried at many places that recognize those factors, has been to institute a regimen of full-blown rigor that makes it quite hard to misrepresent status. But this rigor comes at a high cost.

- **Disadvantage #1—High overhead**: A rigorous system of reporting will have many templates, multiple signoffs, several process rings to jump through, and people to manage these process rings. All this contributes to a high overhead—a significant percentage of resources will be dedicated to oversight rather than to actual throughput.
- **Disadvantage #2—High cost**: Because of the number of people needed to keep a rigorous system in place, cost is likely to be high. This invites justified criticism of resources consumed by excessive overhead that could be better spent on actual projects.
- **Disadvantage #3—Organizational pushback**: A rigorous system that is also resource intensive typically creates resentment among the rank and file. Consequently, adoption is hard and the whole setup is held in place by brute force. This kind of setup also inhibits creativity.
- **Disadvantage #4—Productivity drop**: One of the inevitable disadvantages of a high-rigor performance monitoring system is that it slows everyone down. It robs productivity and diminishes the organization's throughput.
- **Disadvantage #5—Not foolproof**: A subjective system of reporting, while rigorous, can still be manipulated. Projects that are determined to thwart oversight will find ways to misrepresent their status and defeat the whole purpose.

THE ADVANTAGES OF OBJECTIVE INDICATORS

So far we have seen the shortcomings of traditional systems of reporting, which are often subjective and unreliable. Now suppose we were able to come up with an objective system of measuring project performance. What would be the advantages of having such a (as of yet hypothetical) indicator?

- **Advantage #1**: An objective system of measurement would provide credible early warning signs of project trouble, which would be a huge help when it comes to monitoring the project closely. This enhanced monitoring could be instrumental in rescuing the project before too much time and money are wasted.

- **Advantage #2**: An objective system of measurement would enable the portfolio office to take a Pareto approach in managing the portfolio. In other words, the majority of the portfolio office resources could be focused on the (hopefully) few projects that need attention and remediation. At the same time, the majority of projects whose objective measurement indicators show normal performance can be left alone with minimal oversight.
- **Advantage #3**: Objective information allows trend analysis, which helps distinguish between transient noise and serious problems. Almost all projects can show a temporary dip in performance; however, some projects recover soon while others continue to get worse. The presence of objective information allows the portfolio office to distinguish between the two types and focus on the projects that are really in trouble. This enables a far more efficient distribution of the limited resources of the portfolio office.
- **Advantage #4**: As a last resort, underperforming projects may have to be terminated to avoid further wastage of resources. Before that step is taken, there needs to be a solid justification that shows that the termination is warranted. Having objective information provides precisely that justification. It can show that a project has a worsening trend that is unlikely to turn around, leaving termination as the sole option.
- **Advantage #5**: Objective information empowers portfolio governance to make meaningful decisions. In most portfolios, the leadership body or governance is hampered by the availability of good data upon which to make hard decisions. The presence of objective performance data remediates that problem and allows governance to perform their role.

A BASIC EXAMPLE TO INTRODUCE THE CONCEPT OF mEVM

Consider a project to build a freeway with the following specifications:

- Length of freeway = 10 miles
- Duration available = 10 months, starting on January 1
- Budget available = $10M

The mEVM setup is described in the following steps:

- First, there must be a *cost value* declared for each mile of freeway built. In this simplified example, $10M for 10 miles leads to a cost value of $1M for each completed mile.

- Next, there needs to be *schedule value* declared for each mile of freeway built. In this simplified example, 10 months for 10 miles leads to a schedule value of one month for each completed mile.
- Putting the cost value and schedule value together produces the following statement: "Each mile of the freeway will take a month and a million dollars to build."

Using the previously described setup, a baseline now exists to measure the project's progress. After a few months of execution of the road building project, the following progress is observed:

- On May 31, at the conclusion of five months from the start of the project, it is recorded that three miles have been built. The plan stipulates that there needs to be five miles built in five months, since there is a schedule value of one mile per month.
- Also on May 31, it is recorded that the project has spent $6M. The plan stipulates that there needs to be only $3M of spend to build three miles, since there is a cost value of $1M for each completed mile.

Using the previous information, mEVM has the specific answers to the following questions:

- Is the project favorable or unfavorable in cost?
  - Answer: The project is definitely unfavorable in cost, since it has spent $6M to build three miles. The project is spending money at twice the planned rate.
- Is the project favorable or unfavorable in schedule?
  - Answer: The project is definitely unfavorable in schedule, since it has taken five months to build three miles. The project is going much slower than originally estimated.

The potential criticisms leveled at this highway example are that *it's too simple* and also that it's *too linear*. While that's true, this was meant to serve as an introduction only—the intent is to take this technique and apply it to real-life projects in the next chapter. Having looked at the simplest operation of mEVM, here is a brief coverage of the background of this technique and its applicability.

The Origins of mEVM

mEVM is obviously adapted from EVM. The genesis of EVM occurred in industrial manufacturing at the turn of the twentieth century, based largely on the principle of *earned time* popularized by Frank and Lillian Gilbreth, two early advocates of scientific management and pioneers of time and motion study and renowned efficiency experts for their major contributions to the field of industrial engineering.

The concept of EVM became a fundamental approach to project and program management in the 1960s when the U.S. Air Force mandated earned value in conjunction with the other planning and controlling requirements on Air Force programs. The requirement was entitled the Cost/Schedule Planning Control Specification. In 1967, the U.S. Department of Defense (DoD) established a criterion-based approach, using a set of 35 criteria, called the Cost/Schedule Control Systems Criteria (C/SCSC).

In the late 1970s, EVM was introduced to the architecture and engineering industry. David Burstein, a project manager with a national engineering firm is one of the people credited for this, based on his article published in *Public Works Magazine*.

In the late 1980s and early 1990s, EVM emerged as a project management methodology. Why? It is a tool that can provide all levels of management with early visibility into cost and schedule problems. As such, an overview of EVM was included in the Project Management Institute's first *PMBOK® Guide* in 1987, was expanded in subsequent editions, and, more recently, became listed among the general tools and techniques for processes to control project costs.

Over the decades, the concept and its requirements have remained basically unchanged until the late 1990s. It had a few updates to its title such as C/SCSC to EVM Systems Criteria. In the 1990s, many U.S. government regulations were eliminated or streamlined, most notably from 1995 to 1998, EVM criteria was reduced to 32, and ownership was transferred to industry by adoption of ANSI EIA 748-A Standard for EVM Systems.

The U.S. Office of Management and Budget began to mandate the use of EVM across most government agencies, and, for the first time, for certain internally managed projects (not just for contractors). It became a requirement of the U.S. DoD, the National Aeronautics and Space Administration, the Department of Energy, the intelligence community, the Department of Homeland Security, the Federal Aviation Administration, the Department of Transportation, Health and Human Services, and several others.

The construction industry was an early commercial adopter of EVM. Closer integration of EVM with the practice of project management accelerated in the 1990s. EVM received greater attention by publicly traded companies in response to the Sarbanes-Oxley Act of 2002. Today, EVM is used worldwide, but most extensively in the more industrialized parts of the world such as the U.S., Europe, Canada, Australia, China, and Japan.

EVM versus mEVM

One long-standing criticism of EVM is that it needs to gather a ton of data and a large team of people to process it, making it an expensive proposition that only large organizations such as defense companies can afford. Other criticisms of EVM refer to the way it measures the degree of completion of work. These are valid criticisms that need to be addressed. The variant of EVM that we are using in this book has been modified from the original EVM technique to remove the bulk of these criticisms, and the result is called modified EVM or mEVM. mEVM can be considered a *light footprint* variant of EVM that provides most of the benefits of the traditional EVM technique at only a fraction of the cost/effort.

In the next chapter, we'll see the simple building blocks that make up mEVM.

CHAPTER SUMMARY

In this chapter, we took the first step in exploring the powerful technique of mEVM. We covered why traditional solutions are of little use in enabling portfolios to make data-driven decisions and contrasted how mEVM can make a difference. The simple example in this chapter conveyed the basic concept of mEVM, which will be elaborated on in subsequent chapters.

NOTES

A Guide to the Project Management Body of Knowledge. pp. 217–219. Newtown Square, PA: Project Management Institute, 2013.

Defense Systems Management College (1997). *Earned Value Management Textbook.* Chapter 2. Defense Systems Management College, EVM Dept., 9820 Belvoir Road, Fort Belvoir, VA 22060-5565.

Goodpasture, John C. *Quantitative Methods in Project Management.* pp. 173–178. J. Ross Publishing, 2004.

Pisano, Nicholas. "*Technical Performance Measurement, Earned Value, and Risk Management: An Integrated Diagnostic Tool for Program Management.*" Defense Acquisition University Acquisition Research Symposium. 1999.

9

THE ARTIFACTS OF mEVM

INTRODUCTION

The success of any technique or tool ultimately lies in its adoption. In turn, the adoption is driven by the artifacts associated with the process within which that tool or technique functions. What are artifacts? Project management artifacts are tangible by-products produced and used during the execution of projects. Artifacts are either transaction units for a process and/or proxies for project activities and progress. In simpler language, artifacts are the documents we work with to execute a process. If these artifacts are simple to use and few in number, there's a good chance the tool or technique will be widely adopted. In this chapter, we cover the following topics related to modified earned value management (*mEVM*) artifacts:

1. Explore the three main artifacts of mEVM, namely the mEVM graph, the mEVM Excel template, and the mEVM snapshot slide
2. Conduct a detailed walkthrough of each of these artifacts
3. Explore the nuances of using each artifact, including best practices to follow and pitfalls to avoid
4. List the capability maturity levels for these artifacts

THE MAIN ARTIFACTS OF mEVM

The mEVM Graph

Central to the technique of mEVM is the graph that shows both dollars and time in a single graphic—Figure 9.1 shows this graph. Here is an introduction to the features of the graph. The mEVM graph has the following components:

- The horizontal x-axis shows the time progression of the project—typically in months
- The vertical y-axis represents dollars spent, earned, and planned (we'll see shortly what these terms mean)
- The planned value (PV) curve (not shown in the graph) represents the planned path of the project—namely, what is planned to be accomplished at which point in time and at what cost (this will be explored in depth later in this chapter)
- The earned value (EV) curve (not shown in the graph) represents what was actually accomplished by way of deliverables (this too will be explored in depth later in this chapter)
- The actual cost (AC) curve (also not shown in the graph) represents actual dollars spent

The concept behind these curves will be introduced using the same example that was used to introduce the concept of mEVM in the previous chapter. Recall that in the example used in Chapter 8, we were planning to build a 10 mile road in 10 months using a $10M budget. Furthermore, the project started on January 1. The PV curve of the road project would look as shown in Figure 9.2.

The graph basically expects (per the plan) that with every passing month, a mile will be completed at the cost of a million dollars—the graph is cumulative and shows a $10M number at the end of 10 months, with the road being completed.

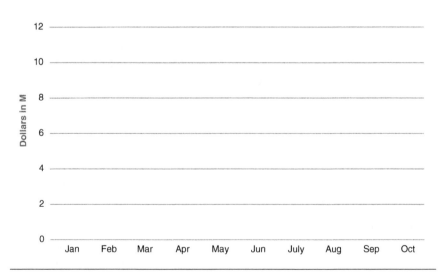

Figure 9.1   Dollars versus time for a road-building project

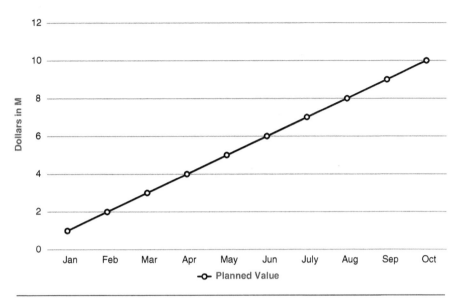

Figure 9.2 The PV curve of the road-building project

Next, let's look at the EV curve—again based on the same example. As we recall, on May 31 (at the conclusion of five months from the start of the project), we saw that three miles had been built. This is shown in Figure 9.3, but some explanation is warranted here. As we calculated at the start of the project, the cost value of each completed mile is $1M. In keeping with this value, the cost value of three completed miles is $3M (notice that we're not talking about the money spent—that will be addressed in the next section).

The last piece of the puzzle is the actual dollar amount spent. Portfolio managers should be keenly aware of the actuals incurred by each project in each month. From the same example used above and in Chapter 8, we report that at the end of May, $6M had been spent. This data is represented graphically in Figure 9.4. Actual dollars spent is probably the simplest of the three curves—it just records the actual spend by month and shows that we have spent a cumulative amount of $6M by the end of May.

By themselves the three curves show valuable data, but what happens if we put the three of them together? Figure 9.5 shows the graph where all three of the curves are put together. This graph is interesting and needs to be explored in detail as outlined here:

- As (1) shows, the PV curve is the *happy path*—the usually optimistic forecast that projects have at the beginning. According to this path, work proceeds smoothly at the planned pace, and nothing costs more than planned.

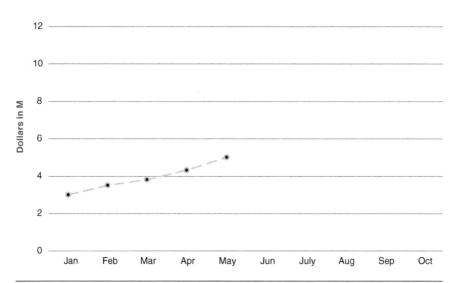

Figure 9.3   The EV curve of the road-building project

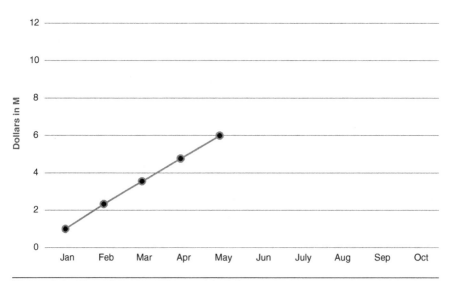

Figure 9.4   The AC curve of the road-building project

- However, as (2) shows, things proceed differently in reality—the progress is modest and the number of miles of highway completed at the end of May is only three as compared to the expected five. Thus, we are lagging behind by two miles.
- Coming to (3), we see that the AC is the money spent in achieving the progress of three miles as described in (2). The AC is $6M as shown in the graph and as described in the example.

There are a couple of standard terms in mEVM that can be described from the graph in Figure 9.5. Schedule variance (SV) indicates how much ahead or behind schedule a project is running. It's basically the distance between the PV curve and the EV curve. In the graph in Figure 9.5, it's (Y coordinate of the EV line at X = May) − (Y coordinate of the PV line at X = May) → 3 − 5 = −2. This is the SV at this point in time (May).

Cost variance (CV) indicates how much over or under budget a project is currently running. It's basically the distance between the AC curve and the EV curve. In this graph, it's (Y coordinate of the EV line at X = May) − (Y coordinate of the AC line at X = May) → 3 − 6 = −3. This is the CV at this point in time (May).

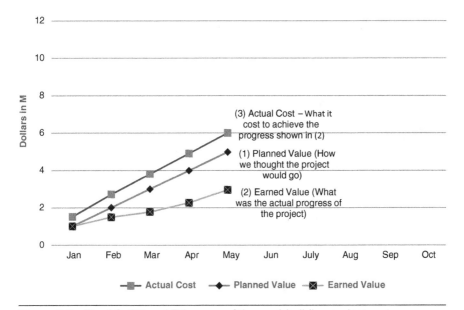

Figure 9.5 The AC, PV, and EV curves of the road-building project

This shows a project in serious trouble—the negative SV shows we have completed less than we expected *and* the negative CV shows that we are also paying more than we expected at the same time.

Why is the mEVM graph so important? Basically, it's a simple, intuitive way to gauge the health of the project. Although the technical indicators such as SV and CV can be calculated from the graph, most people are not interested in those measurements—they are satisfied to know if the project is doing well or not. For that purpose, the graph acts as an easily identifiable *icon* that shows how the project is doing. These three graph lines show a well-performing project, an average-performing project, and a project in distress.

The mEVM Graph of a Project Performing Well

Figure 9.6 shows the mEVM graph that is typical of a well-performing project. The following characteristics are quickly visible in a graph such as this one:

- The EV curve is (for the most part) equal to or higher than the AC curve and the PV curve.
- The cost and the schedule are largely under control right from the beginning of the project.

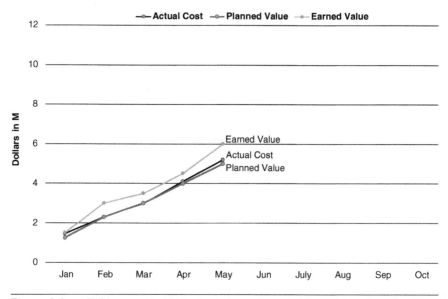

Figure 9.6   mEVM graph that is typical of a project that is performing well

- The three curves are close to each other throughout the duration of the project. Even if they diverge at times, the EV curve is higher than the other two curves, which indicates that the completed work is more than what was planned or budgeted.

The mEVM Graph of a Project with an Average Performance

Figure 9.7 shows the mEVM graph that is typical of a project with an average performance. The following characteristics are quickly visible in a graph such as this one:

- The EV curve is, at least in some places, equal to or higher than the AC curve and the PV curve.
- The cost and the schedule are largely under control right from the beginning of the project.
- The three curves are close to each other throughout the duration of the project. If there is a gap between the three, it is likely to be small and does not get bigger with each reporting cycle. That is, the project fundamentals do not show a worsening trend.

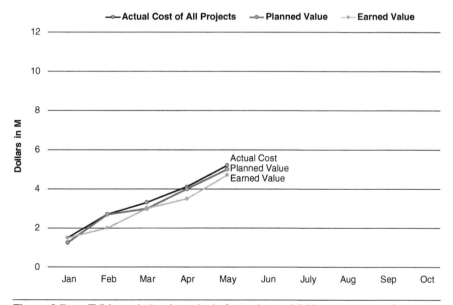

Figure 9.7   mEVM graph that is typical of a project exhibiting average performance

The mEVM Graph of a Project in Crisis

Figure 9.8 shows the mEVM graph that is typical of a project in crisis. The following characteristics are quickly visible in a graph such as this one:

- The EV curve almost always lags the AC curve and the PV curve. This trend is established early in the project and either sustains or becomes more pronounced as the project continues.
- The position of the EV curve lagging the other two indicates two things—deliverables take longer than anticipated and, at the same time, these deliverables also cost more than anticipated.
- The cost and the schedule are out of control right from the beginning of the project.
- The three curves are not close to each other and only diverge further throughout the duration of the project.

Experience on numerous engagements has shown the graph to be valuable in showcasing true project status to senior stakeholders, including governance and executive management. The placing of the three curves on the mEVM graph provide an intuitive way to grasp what is happening in the project, while backed by quantitative data. We'll revisit the mEVM graph in more detail in later chapters.

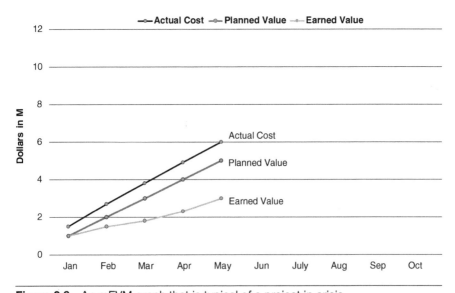

Figure 9.8   An mEVM graph that is typical of a project in crisis

The mEVM Excel Template

As covered in the previous section, the mEVM graph is the most recognizable artifact of the mEVM system. However, the data that is represented in the graph comes from another artifact, simply called the mEVM Excel template. This section will deal with the details of that template and how to use it successfully.

Figure 9.9 shows an example of the mEVM template for the project titled Implement Payment Gateway Interface Project. The following are the key features of this template:

1. Column A is the list of deliverables that are produced during the life of the project
2. Column B is the EV column—it is a calculated column and the details behind the calculation are explained later in this chapter
3. Columns C and D are the start and end dates for each deliverable listed in column A
4. Column E is a status column that can be set only to one of the following three states: *not started*, *started*, or *complete*
5. Columns F through Q represent the months of the year and contain the dollar cost estimate of the corresponding deliverable
6. Column R is the total of columns F through Q

Each of the columns in Figure 9.9 needs to be looked at in detail to understand how the overall template works.

Column A—The Deliverable

This column contains the deliverables that need to be completed as part of the project. The deliverables need to meet the following characteristics in order to be effective in tracking performance:

- **Characteristic #1**: The statement of what will be delivered should be unambiguous and concrete in nature—there should be no leeway for confusion or misinterpretation.
  - Example: *Approval of the overall outline of the information technology (IT) project plan by stakeholders* is very straightforward in that it stipulates not only the creation of the IT project plan, but also approval of the same by the stakeholders.
  - Counterexample: *Create overall outline of the IT plan* is ambiguous and not concrete enough. Does it refer to the first draft of a plan? Has it been socialized? What if some of the stakeholders object to the plan? How will it be revised? These are some of the questions still outstanding.

	A	B	C	D	E	F	G	H	I	J	K	L		M	N	O	P Q	R
1	**Implement Payment Gateway Interface Project - Track Subdeliverables for mEVM attainment**																	
2	**Project Work Packages and Deliverables**	**Current EV**	**Start Date**	**End Date**	**Status**	**Jan**	**Feb**	**Mar**	**Apr**	**May**	**June**	**July**	**Aug**	**Sept**	**Oct**	**Nov**	**Dec**	**Total $'s**
3	**Development of IT Project Workstream Plan**		**1/2/2017**	**4/30/2017**														
4	Overall outline of IT Project Plan	**$0.00**	1/2/2017	1/31/2017	**Open**	$20,000												$ 20,000
5	Selection of Payment Gateway Implementation vendors	**$0.00**	2/1/2017	2/28/2017	**Open**		$15,000											$ 15,000
6	User Identity Validation Implementation	**$0.00**	3/1/2017	3/31/2017	**Open**			$25,000										$ 25,000
7	Logical Access Implementation	**$0.00**	4/1/2017	4/30/2017	**Open**				$15,000									$ 15,000
8	**Solution Design Document Completion**		5/1/2017	7/31/2017														$ -
9	Logical Solution Design Completion	**$0.00**	5/1/2017	5/31/2017	**Open**					$10,000								$ 10,000
10	Code Development Design Completion	**$0.00**	5/1/2017	6/30/2017	**Open**						$15,000							$ 15,000
11	Architecture Solution Strategy Completion	**$0.00**	6/1/2017	7/31/2017	**Open**							$25,000						$ 25,000
12	Environment implemented	**$0.00**	7/1/2017	7/31/2017	**Open**							$20,000						$ 20,000
13	Totals	**$0.00**				$20,000	$ 15,000	$25,000	$ 15,000	$ 10,000	$ 15,000	$45,000						$ 145,000

Figure 9.9 An example of the mEVM template for Implement Payment Gateway Interface Project

- **Characteristic #2**: Deliverable completion status should be digital—either it is complete or it isn't. There should not be another possibility.
  - Example: *Prototype website design complete and approved by customer* is an example of a deliverable that is easy to classify as complete or incomplete.
  - Counterexample: *Website load testing* is an activity that is hard to pin down as complete or still in progress.
- **Characteristic #3**: Deliverables need to be outcome based, not activity based—the deliverable should indicate something tangible that can be verifiably delivered. The deliverables should not just be activity performed.
  - Example: *Website updated with new customer facing content* is a tangible, impactful outcome.
  - Counterexample: *Meetings to brainstorm new customer facing content* is an activity that may or may not lead to a tangible outcome. Success lies in focusing on outcomes rather than on intermediary activities.

Taken together, the characteristics result in the selection of only the most essential, critical sub-products that chart the progress of the project from the beginning to the end. In other words, these are the kind of *deliverables/steps* whose successful completion indicates successful execution of the project. Conversely, if these steps are not getting done on time or on budget, it is highly improbable that the project as a whole is on time or on budget. A few other pointers to choosing the best deliverables are:

- **Right sized deliverables**—Deliverables should strike the right balance between too big and too small. Huge milestones create lag, distort the graph, and make it choppy; whereas small milestones create a lot of overhead in tracking. A good rule of thumb is to avoid picking any milestone bigger than 5% of the project budget (exceptions with rationale permitted). Large milestones can be broken into smaller pieces but must still adhere to the milestone characteristics that were previously described.
- **Right frequency**—This is related to right sizing deliverables. By choosing the right frequency of deliverables completion, the project can have an even flow of completion measurement, rather than large numbers of deliverables clumped together in the same months and having no deliverables in other months. A good rule of thumb is to have at least one deliverable planned to complete in each month.
- **Deliverable effort calculation**—Each deliverable, once chosen, needs to have a start date and an end date. Then comes an estimation of what resources would be needed to complete the deliverables and how much

these resources cost, leading to a fairly straightforward calculation of what dollar amount of effort it would take to complete that deliverable.

Column B—The EV Column

This column calculates the EV of the project at any given time. Each field in this column contains a simple formula that works as follows:

- If the deliverable is complete, grant the full PV of the deliverable as *earned.*
- If the deliverable has started but is not yet complete, grant 25% of the PV of the deliverable as *earned*—the other 75% of the PV will be earned when the deliverable is complete.
- This 25/75 split in granting EV is called an *earning rule.* Other earning rules are possible—0/100, 50/50, 75/25, etc. For example, the 0/100 earning rule provides no EV until the deliverable is complete. Each earning rule comes with its own pros and cons—it is a design decision as to which earning rule to use.

A cell at the bottom of this column sums all of the EV of the project at any given time and this forms the EV number of the project that is then plotted on the graph.

Columns C and D—The Start and End Dates

Column C shows the start date of the work that results in a deliverable. For example, if a months-long effort is started on February 1 to ultimately create a deliverable called *website design complete,* the start date for that deliverable is 2/1. If the months-long effort is planned to complete at the end of July, the finish date is 7/31.

The start and end dates can be used to perform a light kind of tracking as follows: consider the *website design complete* deliverable, originally planned to start on 2/1. When it appears that this cannot start on 2/1 due to a delay in some preceding dependency, it's a good idea to update the task to show the new date along with notes showing that the date had been changed and why it had to be changed. This could repeat a few more times (since this template is reviewed every month).

Similarly, for the end date, deliverables that do not complete on time are noted as *late,* along with a new date and a reason why they were delayed. These notes are also valuable in providing commentary in explaining variance when showing the graph to governance and other stakeholders. The notes can be inserted within the Excel cell using the *create note* feature—this way all the information is available on the template while preserving its compact design.

Column E—The Status of the Deliverable

Each deliverable can only have one of the following status indicators:

- Not yet started
- Started, but not yet finished
- Finished

It's very important to adhere to only these three status choices, because Column B contains a formula that parses this column to decide what the EV should be. Any value in this cell that is not recognized by the formula can give rise to erroneous results. It may be preferable to have a drop down (instead of free text entry) for every cell in this column, so that data entry errors can be avoided.

Columns F through Q—Months of the Year

These columns represent the months of the year from January to December. If a deliverable whose effort is estimated at $10,000 is planned to be finished in the month of March, there would be a $10,000 entry in the March column. It does not matter if the deliverable started in the month of January or February, the full dollar estimate of the effort needs to be in the month of planned completion—which is March in this example.

More than one deliverable can complete in any given month, so there may be multiple entries in any monthly column. However, there should only be one entry in any row. Another point to ponder—what happens (in this example) if March is over but the deliverable is not yet complete? The answer to that is that the dollar amount stays in the March column.

Column R

Column R holds the total of the columns from F to Q. Its main function is to provide an intermediate total to be used in the formula to populate Column B.

How to Use the mEVM Excel Template

The Excel template is created at the beginning of the project, but is typically updated every month and the output is fed into the graph. Here are the steps involved in the one-time setup and in the monthly update:

One-time Setup

- **Step 1**: All of the deliverables of the project need to be drawn up in the Excel format that was described earlier, along with start and end dates. Care must be taken to ensure that each deliverable meets the guidelines described in the previous section.

- **Step 2**: The effort needed to complete each deliverable needs to be converted to a dollar amount that is based on the resources involved in that effort; and the final dollar amount needs to be inserted in the column corresponding to the month of the completion of the deliverable.
- **Step 3**: A monthly column may have more than one entry. All the entries in a column need to be added to calculate totals for each month—this total constitutes the PV for that month.
- **Step 4**: The PV for each month is entered into a table on the second tab of the Excel sheet, as shown in Table 9.1.

The information contained in Table 9.1 will be used to create the mEVM graph. To start with, it would only show the PV curve, since the data areas for the other two curves (AC and EV) are blank in the table. More details can be found in the previous section in this chapter that deals with the mEVM graph.

Ongoing Monthly Update

Every month, the Excel template has to be updated according to the steps outlined here:

- **Step 1**: The list of deliverables is reviewed to see if any new effort was started. If so, the *status* of that deliverable should be changed to *in progress*.
- **Step 2**: The list of deliverables is reviewed to see if any previously started effort is now complete. If so, the *status* of that deliverable should be changed to *complete*.
- **Step 3**: If a deliverable was supposed to complete, but the completion date has now been pushed back, the finish date should be updated to show the new date. A comment should be inserted in that cell using the Excel *note* feature, for tracking purposes.
- **Step 4**: As a result of all of these status changes, the EV entries may change (calculated by a formula embedded in the cells of Column B).
- **Step 5**: The total EV, as noted in the totals cell under Column B, is transferred to the table in the second tab of the Excel template, in the EV row, for the corresponding month (say, in January).
- **Step 6**: The AC incurred by the project that month is entered in the table in the second tab of the Excel template, in the AC row, for the same month (January). The complete table looks as shown in Table 9.2.

This data is then used to update the graph. (Note: the data as shown in the graph is cumulative—meaning that each month's data is added to the sum of all the previous months' data before being shown as a point on the graph.)

Table 9.1 Table to make cumulative entries for the PV, EV, and AC for the project each month

	Jan	Feb	Mar	Apr	May	June	July	Aug	Sept	Oct	Nov	Dec
Planned Value												
Earned Value												
Actual Cost												

Table 9.2 Table showing the PV, EV, and AC for the project for the month of January

	Jan	Feb	Mar	Apr	May	June	July	Aug	Sept	Oct	Nov	Dec
Planned Value	10,000											
Earned Value	9,000											
Actual Cost	11,000											

The mEVM Snapshot Slide

Although we saw the details of the graph and the Excel template, it must be kept in mind that these are the *details* that the executive stakeholders may not be interested in. Senior stakeholders are often looking for only a simple, concise snapshot that shows *how the project is performing* and a brief explanation of *why the project is performing in this way*. This information is captured in the mEVM snapshot slide. Often, it is the only thing that will be shown at portfolio governance meetings. Figure 9.10 shows what the mEVM snapshot slide would look like for the sample project XYZ for the month of December. The major components of the mEVM snapshot slide are explained here:

1. The mEVM graph shows the latest updated EV and AC numbers and compares them to the PV curve
2. The table to the right shows the two most important metrics—CV and SV as of this month (December)
3. The executive takeaway box at the bottom provides the concise commentary that executives are looking for

Project XYZ – December EVM Snapshot

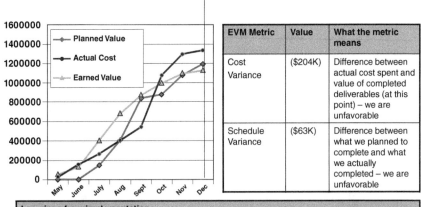

Figure 9.10   December mEVM snapshot slide for Project XYZ

Ongoing Monthly Update

The snapshot slide is created every month and shows the latest data for that month. For example, Figure 9.10 showed the data for December. It is recommended to keep all the previous monthly slides for that project in one PowerPoint file. This aids trackability for executives in case there is a request to revisit the performance information from previous months.

Levels of Portfolio Capability Maturity

Level 1

- The organization has just begun to adopt mEVM. Consequently, there is little standardization by way of common templates that all projects can adopt.
- Without templates, the organization finds it hard to ensure adoption by the rank and file.
- Without standard templates, every project performs mEVM differently, with some of them practicing the technique incorrectly.
- Due to the variability caused by the absence of templates, the mEVM review is contentious and portfolio governance is not able to obtain many of the promised benefits of mEVM.

Level 2

- The organization is further along in its mEVM journey and has begun to roll out common templates that all projects can adopt.
- Although the mEVM templates are being adopted, there is not enough uniformity in their usage among the different projects. Some possible situations are described here:
    - In the case of the mEVM Excel template, some projects may not follow sufficient rigor in ensuring that the project deliverables are definitive and concrete in nature.
    - Some projects may not apply rigor in selecting milestones with digital completion status. Also, such projects may be non-exacting in what constitutes 100% completion of a deliverable.
    - Some projects may not adhere strictly to earning rules in granting EV to project activities. Consequently, this may lead to a non-uniformity in comparing project performance.
- Due to the aforementioned lack of rigor, portfolio governance may have mixed confidence in the data that they are reviewing.

- The organization may also not be sufficiently organized in the storage and retrieval of mEVM templates for previous reporting periods.

Level 3

- The organization has attained maturity in its mEVM journey and has rolled out common mEVM templates for all projects.
- All of the mEVM templates have been made mandatory in order to qualify for portfolio funding; and this ensures complete adoption. Furthermore, the portfolio office serves as a quality control for ensuring that all projects are using mEVM with the appropriate rigor as evidenced by the circumstances described here:
  - In the case of the mEVM Excel template, all projects are made to select project deliverables that are definitive and concrete in nature.
  - All projects are allowed to only select milestones with digital completion status. Furthermore, the portfolio office ensures that all projects adhere to a strict standard in defining what constitutes 100% completion of a deliverable.
  - All projects follow a common standard for applying earning rules while granting EV to project activities. This ensures uniformity when comparing project performances.
- Due to the enforced rigor, portfolio governance has confidence in the data that they are asked to review.
- The organization has streamlined the storage and retrieval of mEVM templates for previous reporting periods. This ensures easy access to historical information as well as re-use of information—because it is far easier to edit existing information than to recreate new information from scratch.

CHAPTER SUMMARY

In this chapter, we dealt with the operational artifacts of mEVM. As part of that treatment, we detailed the three important artifacts of mEVM: the mEVM graph, the Excel template, and the snapshot that displays mEVM information to the executives in a concise manner. We approached the nuances involved in the creation and monthly updates for each of these three templates. For the mEVM graph, we assembled the graph using the simple example of a roadbuilding project and went on to show the typical mEVM graphs associated with whether the projects are performing well, average, or poorly. We then performed a walk-through of the mEVM Excel template, including a detailed look at each of the columns and recommended best practices for each of the columns. Finally, we covered the need for the mEVM snapshot template and explained how it addresses the status need for executives. The chapter concluded with the levels of maturity for this portfolio capability.

NOTE

1. Wikipedia, *The Page on Artifacts*, https://en.wikipedia.org/wiki/Artifact _(software_development)

10

AGGREGATING mEVM DATA FOR THE ENTERPRISE

INTRODUCTION

In Chapter 9 we covered in detail the modified earned value management (*mEVM*) artifacts for a project. The mEVM artifacts enable the accurate tracking and forecast of spend for each project. While each project is important, the strength of a portfolio lies in its power to see the forest as compared to individual trees. This chapter introduces the concept of aggregating the individual project data and explores the following aspects of aggregation:

1. Explore the drivers of aggregation
2. Explore the dimensions of aggregation
3. Discuss how to aggregate mEVM data to create and maintain the portfolio view
4. Discuss how to aggregate mEVM data to create and maintain the program view
5. Discuss how to aggregate mEVM data to create and maintain the multi-year project view
6. Discuss how to aggregate mEVM data to create and maintain the multi-year program view
7. Discuss how to aggregate mEVM data to create and maintain the multi-year portfolio view
8. Explore how aggregated data is consumed by portfolio governance
9. List and elaborate on the success factors of portfolio data aggregation

THE DRIVERS OF AGGREGATION

- **Need #1—Switching between the big picture and the details**: The portfolio operates at the aggregate level and this necessitates the creation of the big picture by aggregating all the component data. Conversely, the big picture needs to be able to drill down, if appropriate, to the individual units of work (projects).
- **Need #2—Optimizing the use of governance time**: Portfolio governance is made up of senior leaders whose time is typically precious. It is inefficient to provide detailed portfolio data to these leaders, who have neither the time nor the patience to wade through this mass of data. Therefore, the data will need to be aggregated so that it can be consumed at a summary level by senior leaders. At the same time, the option of diving into the details must be preserved.
- **Need #3—Visualize, analyze, and predict the trend**: There is a need to observe how the portfolio is doing over time. To create this trend line, there needs to be an aggregation of data from the individual efforts, namely the projects. It's much easier to represent the progression over time of a single, aggregated quantity than it is to show the progression of a collection of numbers.

THE THREE DIMENSIONS OF AGGREGATION

While there are many ways of recombining data, here are the three types of aggregation that find most applicability in a portfolio.

- **Dimension #1—The portfolio view**: The portfolio view is the most essential aggregation, namely, displaying the data of all the projects in the portfolio in one view.
- **Dimension #2—The program view**: The program view aggregates the data of all the projects contained within the program and displays the same.
- **Dimension #3—The multi-year project view**: The multi-year project view aggregates all the data of a project, spread out over multiple years, and displays the same in one integrated view.

THE PORTFOLIO VIEW

This simple view shows all of the mEVM data of the projects in the portfolio in a single table, as shown in Table 10.1. Notice that this table shows only the monthly status, that is, everything here is relevant as of that month only, and

Table 10.1 Monthly portfolio dashboard

Name of the Project	Total Budget of the Project	How Much Money Did We Actually Spend?	How Much Work Should Have Been Completed?	How Much Work Did We Actually Complete?	What Is the Cost Efficiency?	What Is the Schedule Efficiency?
(A)	(B)	(C)	(D)	(E)	(F)	(G)
Project 1	$500K	$100K	$100K	$75K	0.75	0.75
Project 2	$750K	$300K	$200K	$300K	1	1.5
Project 3	$100K	$50K	$40K	$25K	0.5	0.625
Project 4	$240K	$200K	$175K	$175K	0.875	1
Project 5	$1,000K	$500K	$450K	$500K	1	1.11
...	...	...	...	...	...	...
...	...	...	...	...	...	...

will be updated for the next month. All the data entries in the table are for illustration purposes only.

Key Elements of the Portfolio View

The following is a list of the key elements of the portfolio view:

- **Column A—Project name**: This column contains the names of the projects in the portfolio. The name of each project is hyperlinked to the detailed mEVM data of that project. For example, by clicking on Project 1, the user would go to a web page containing the mEVM snapshot of Project 1 showing the mEVM graph and other details, as explained in Chapter 8.
- **Column B—Project budget**: This column contains the official budget of each project in the portfolio. For example, Project 1 has a total budget of $500K.
- **Column C—How much money was actually spent**: This column contains the actual amount of money spent by each project. This is the same as the actual cost (AC) from the mEVM data. For example, Project 1 has spent $100K as of July.
- **Column D—How much work should have been completed**: This column specifies how much work should have been done at this point in time in the project's timeline. This is the same as planned value (PV) from the mEVM data. For example, Project 1 should have finished deliverables worth a cumulative amount of $100K as of July.
- **Column E—How much work was actually completed**: This column specifies the value of work that was actually completed. As described in Chapter 8, mEVM allots a specific amount of dollars for each deliverable. When those deliverables are complete, the project is allowed to *recognize* the value of the completed work. For example, Project 1 has finished deliverables worth a cumulative amount of $75K as of July. This is the value of work that has actually been completed.
- **Column F—What is the cost efficiency**: This column measures the ratio of Column E to Column C—the ratio of completed work to the money being spent. Cost efficiency is just another term for cost performance indicator from mEVM data. For example, Project 1 has finished deliverables worth a cumulative amount of $75K against an actual spend of $100K, creating a cost efficiency number of 0.75.
- **Column G—What is the schedule efficiency**: This column measures the ratio of Column E to Column D; in other words, the ratio of completed work to the work that was planned to complete at this point in time. Schedule efficiency is just another term for schedule performance

indicator from mEVM data. For example, Project 1 has finished deliverables worth a cumulative amount of $75K against a planned amount of $100K, creating a schedule efficiency number of 0.75.

How to Create the Portfolio View

Here are the steps for creating the portfolio view:

- **Step 1**: The structure of Table 10.1 is created with Columns A through G, as shown.
- **Step 2**: Column A is populated with the list of projects in the portfolio, with each project on a separate row. Columns B through G are populated with the current mEVM data for that month.
- **Step 3**: The name of the project in Column A is a hyperlink to the detailed mEVM data for that project including the mEVM graph and the mEVM Excel template.
- **Step 4**: The completed table is included in the governance presentation package for the month.

How to Operate the Portfolio View

Here are the steps necessary to operate the portfolio view:

- **Step 1**: At the (monthly) governance session, this table is shown to governance members. It's worth recalling here that during the roll out of mEVM to the enterprise, the governance team needs to be trained on how to understand the mEVM data being presented to them. This table is one of the artifacts to be included in that training.
- **Step 2**: Portfolio governance would typically want to focus on underperforming projects. These are projects with a low score in Column F (cost performance) and Column G (schedule performance). In the beginning stages of mEVM adoption, the portfolio manager may have to recommend that the governance team take a closer look at underperforming projects. Over the course of time, portfolio governance begins to understand how the mEVM system works and knows which projects to focus on.
- **Step 3**: Column A contains the hyperlinked names of the projects. If a project's name is clicked on, it brings up the mEVM data for that project to enable portfolio governance to take a closer look at the project's details. For example, Project 3 in Table 10.1 seems to have low scores in cost and schedule efficiency. When the hyperlinked title of Project 3 (found in Column A) is clicked on, it takes the user to the mEVM graph and other mEVM artifacts of Project 3. These are then browsed upon by

portfolio governance to arrive at a decision for Project 3. This process is repeated for all the underperforming projects that the portfolio governance wishes to review.

- **Step 4**: Portfolio governance needs to follow one of the following options on each underperforming project:
  - Wait and watch—wait for a more pronounced trend in the project before taking any action
  - Put the project on the enhanced monitoring list (EML)
  - If the project is already on the EML, decide to either keep it there for further observation or terminate the project
- **Step 5**: Review concludes for the current portfolio cycle

All of these steps are meant for one portfolio cycle, which is assumed to be monthly in the previous example. For the next portfolio cycle, the portfolio view will need to be recreated with current mEVM data (mEVM data that is current at the next portfolio cycle), and the process continues.

THE PROGRAM VIEW

A program is a collection of related projects. A portfolio contains not only individual projects but also programs that, in turn, are composed of related projects. What would a program view aggregation look like? Table 10.2 shows a program view for Program ABC composed of the constituent projects.

It needs to be reiterated that only the projects belonging to Program ABC should be in that program's view. When Columns B through E are aggregated, the program totals are produced. The program total for cost efficiency (or, program cost efficiency) is calculated as follows:

$$\text{Program cost efficiency} = \left(\frac{\text{Sum of actually completed work across all the projects}}{\text{Sum of actual money spent across all the projects}} \right)$$

In this case, program cost efficiency = $1,075K/$1,150K = 0.93.

Similarly, program schedule efficiency is calculated as follows:

$$\text{Program schedule efficiency} = \left(\frac{\text{Sum of actually completed work across all the projects}}{\text{Sum of planned work across all the projects}} \right)$$

In this case, program schedule efficiency = $1,075K/$965K = 1.11.

Connecting the Program View to the Portfolio View

A program is a composite of related projects. By aggregating all of the mEVM data of the projects belonging to Program ABC, we obtained the *program total row* that is at the bottom of Table 10.2. How would it look if we included the

Table 10.2 Monthly program view of Program ABC showing performance of projects within the program

Name of the Project (A)	Total Budget of the Project (B)	Actual Money Spent (YTD) (C)	Planned Work (YTD) (D)	Earned Value (YTD) (E)	Cost Efficiency (F)	Schedule Efficiency (G)
Project 1	$100K	$50K	$40K	$25K	0.5	0.625
Project 2	$240K	$200K	$175K	$175K	0.875	1
Project 3	$1,000K	$500K	$450K	$500K	1	1.11
Project 4	$500K	$100K	$100K	$75K	0.75	0.75
Project 5	$750K	$300K	$200K	$300K	1	1.5
Program ABC's Total	$2,590K	$1,150K	$965K	$1,075K	0.93	1.11

program total row in the portfolio view? The result is shown in Table 10.3, in a portfolio view containing both projects and programs.

The first row of Table 10.3 shows Program ABC and its mEVM data, which is the data shown in the program total row. Remember that this data is obtained by adding the mEVM data of the corresponding sub-projects in the program, as shown in Table 10.2.

As a best practice, the entry *Program ABC* should be hyperlinked to Table 10.2. It would help in the following way: portfolio governance would review Table 10.3 as part of the portfolio review and view the mEVM data of Program ABC to get an idea of how the program is performing on the whole. If a deeper view is desired, Program ABC can be clicked on and the embedded hyperlink will bring up Table 10.2 that shows how each of the sub-projects within Program ABC are functioning.

How to Create the Program View

- **Step 1**: The program view table for Program ABC is created as shown in Table 10.2. The steps to create this view are identical to the steps used to create the portfolio view (Table 10.1). The one difference is that only the projects which belong to the program are included in the program view table (as opposed to the portfolio view, where all the projects in the portfolio are included).
- **Step 2**: The mEVM data of all the rows are summed up and the total is recorded at the bottom row. This is the Program Total for mEVM.
- **Step 3**: A row is created for Program ABC in Table 10.3 (the portfolio view). In this row, the Program Total numbers from Table 10.2 are recorded. The name of the program in Table 10.3 (i.e. "Program ABC") is hyperlinked back to Table 10.2.

Steps 1 through 3 represent the actions for one reporting cycle. The following are the steps to update the program view for the next reporting cycle:

- The program view table for this program needs to be recreated with the new mEVM entries for the sub-projects for the current month.
- The new program total needs to be calculated.
- A new row needs to be created for this program in the new portfolio view that has been created for the current month; the updated program total needs to be entered in this new row.
- As mentioned before, a hyperlink needs to connect the program entry on the portfolio view to the program view table.

Table 10.3 Portfolio view showing performance of both programs and individual projects in a given year

Name of the Project/ Program	Total Budget of the Project/ Program	How Much Money Did We Actually Spend?	How Much Work Should Have Been Completed?	How Much Work Did We Actually Complete?	What Is the Cost Efficiency?	What Is the Schedule Efficiency?
(A)	(B)	(C)	(D)	(E)	(F)	(G)
Program ABC	$2,590K	$1,150K	$965K	$1,075K	0.93	1.11
Project A	$100K	$50K	$40K	$25K	0.5	0.625
Project B	$240K	$200K	$175K	$175K	0.875	1
Project C	$1,000K	$500K	$450K	$500K	1	1.11
Project D	$500K	$100K	$100K	$75K	0.75	0.75
Project E	$750K	$300K	$200K	$300K	1	1.5
Total	**$5,180K**	**$2,300K**	**$1,930K**	**$2,150K**	**$0.84**	**$1.02**

THE MULTI-YEAR PROJECT VIEW

A common problem associated with annual cycles of funding is that it hampers the visibility of performance associated with multi-year projects. It is not easy to keep track of how a multi-year project has performed in all the different years of its existence. However, with the adoption of mEVM, there is an objective way to capture and represent the multi-year performance of a long-running project. This method is explained with the information shown in Table 10.4.

Consider a project (Project A) that had a total budget of $100K for Year 1 (Column F shows the year). At the end of Year 1, we see the mEVM data for the project as shown in Columns B through E. Imagine that the project was extended for another year (Year 2)—again with a total budget of $100K. The project executes through Year 2 and at the end of the year, the historical mEVM data for both Year 1 and Year 2 are shown in Table 10.5.

If this project continues in a similar way for several years, the table representing all the historical mEVM data over the years would look like Table 10.6—the multi-year view of Project A.

Table 10.6 has two additional columns that have not been seen previously—a cost efficiency column and a schedule efficiency column for each year. Another enhancement is that hyperlinks have been embedded into each of the years in Column F in Table 10.6. When clicked on, these hyperlinks would take the user to the mEVM artifacts of the project for that specific year so that a more detailed look can be obtained about the performance of the project for that year. Table 10.6 also has a totals row at the bottom that provides a quick look at the total figures across all years.

This table is, by itself, a huge improvement over what may have been provided in the absence of mEVM. At a glance, it not only shows how much money was allotted to the project in total and in each of the various years, but also provides the much needed insight of what scope was actually achieved and how much was actually spent. However, the insight provided by the table becomes apparent when translated into a graph, as shown in Figure 10.1.

The most visible takeaway from this graph is that while Project A has delivered all the scope elements according to plan (as evidenced by the earned value (EV) curve coinciding with the PV curve), there has been a significant cost overrun occurring through the years (as evidenced by the AC curve being significantly higher than both the EV and the PV). This insight is hidden and is probably not easily obtainable for multi-year projects at most organizations. While this lack of visibility helps the project owners hide the unflattering performance of such projects, it puts the portfolio office at a distinct disadvantage. The multi-year view remediates this disadvantage.

Table 10.4 mEVM data for Project A in Year 1

Name of the Project	Total Budget of the Project	How Much Money Did We Actually Spend?	How Much Work Should Have Been Completed?	How Much Work Did We Actually Complete?	Year of Operation	What Is the Cost Efficiency?	What Is the Schedule Efficiency?
(A)	(B)	(C)	(D)	(E)	(F)	(G)	(H)
Project A	$100K	$90K	$100K	$95K	Year 1	1.06	0.95

Table 10.5 mEVM data for Project A in Year 1 and Year 2

Name of the Project	Total Budget of the Project	How Much Money Did We Actually Spend?	How Much Work Should Have Been Completed?	How Much Work Did We Actually Complete?	Year of Operation	What Is the Cost Efficiency?	What Is the Schedule Efficiency?
(A)	(B)	(C)	(D)	(E)	(F)	(G)	(H)
Project A	$100K	$90K	$100K	$ 95K	Year 1	1.06	0.95
	$100K	$120K	$100K	$100K	Year 2	1.00	1.00

Table 10.6 Multi-year view of Project A's performance

Name of the Project (A)	Total Budget of the Project (B)	How Much Money Did We Actually Spend? (C)	How Much Work Should Have Been Completed? (D)	How Much Work Did We Actually Complete? (E)	Year of Operation (F)	What Is the Cost Efficiency? (G)	What Is the Schedule Efficiency? (H)
Project A	$100K	$90K	$100K	$95K	Year 1	1.06	0.95
	$100K	$120K	$100K	$100K	Year 2	0.93	0.98
	$125K	$150K	$125K	$135K	Year 3	0.92	1.02
	$150K	$200K	$150K	$140K	Year 4	0.84	0.99
	$50K	$100K	$50K	$60K	Year 5	0.80	1.01
Multi-Year Project Total	$525K	$660K	$525K	$530K		0.86	1.00
						...	...

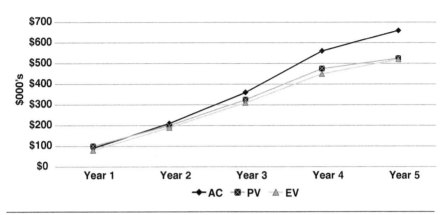

Figure 10.1   Multi-year performance graph of Project A

How to Create the Multi-Year Project Graph

This graph is constructed by plotting the cumulative data for the PV, EV, and AC of the project for each year. The data from Table 10.4 is accumulated for every year as shown in Table 10.6. Each of the curves is then plotted one by one. For example, the AC (see Column C in Table 10.6) was $90K in Year 1, $120K in Year 2, $150K in Year 3 and so on. Plotting these points on the graph and connecting them gives us the AC curve seen in Figure 10.1.

Similarly, the PV (see Column D in Table 10.6) was $100K in Year 1, $100K in Year 2, $125K in Year 3, and so on. Plotting these points on the graph and connecting them gives us the PV curve seen in Figure 10.1.

Finally, the EV (see Column E in Table 10.6) was $95K in Year 1, $100K in Year 2, $135K in Year 3, and so on. Plotting these points on the graph and connecting them gives us the EV curve seen in Figure 10.1.

When these three curves are plotted, the complete visual representation of the project's multi-year journey is obtained as seen in Figure 10.1. The above procedure needs to be done for every project that has a multi-year journey, in order to generate a graph for each project.

Applications of Multi-Year Project View

- With the help of aggregated mEVM data, a project's multi-year performance can no longer be masked. The true performance of the project in each year, as well as the cumulative performance over all the years, becomes visible.
- The multi-year view is an important piece of data during annual planning decisions that seek to approve or deny the continued execution of multi-year projects.

- During portfolio balancing exercises, the multi-year views of all long-running projects can be examined to determine which projects are *losing bets* and should be candidates for elimination.
- The enhanced visibility into the performance of multi-year projects creates more accountability in the sense that multi-year projects cannot expect to automatically be funded in additional years. This funding decision is now a function of whether the performance over the years has been on par with the other projects.

THE MULTI-YEAR PROGRAM VIEW

The multi-year program view combines the concepts of the program view and the multi-year project view that was just explored at length. Table 10.7 shows what a multi-year program view would look like.

How to Create the Multi-Year Program View

1. Each row in the multi-year program view in Table 10.7 is obtained by summing up the projects in that program for that year. For example, the totals row in Table 10.2 would become the Year 1 (or Year 2, or Year 3, etc.) row for Table 10.7.
2. It's highly recommended to embed a hyperlink in the *year* entry in each row of Table 10.7. This hyperlink would point back to the single-year program view (Table 10.2), whose total is being used as a row in Table 10.7.
3. Unlike a project, a program's composition could vary over the years. For example, in Year 1, it could have five sub-projects, while in Year 2 it may have seven sub-projects. This underscores why it's a good idea for the hyperlink to point to a detailed view that shows what projects were included in the program for that year. This is also a good reason to keep accurate historical records.

The Multi-Year Program View Rendered as a Graph

Although the multi-year program view is useful in a table, as shown in Table 10.7, it's even more insightful to render the multi-year program information into a graph because that provides a more intuitive view, especially for portfolio governance (see later section in this chapter regarding presentation of aggregated data to portfolio governance). Since a program is usually more than a year long, the graph artifact is a good fit for providing visibility into the program's multi-year performance. One of the readily apparent insights from the previous

Table 10.7 Multi-year view of Program ABC's performance

Name of the Program (A)	Total Budget of the Project (B)	How Much Money Did We Actually Spend? (C)	How Much Work Should Have Been Completed? (D)	How Much Work Did We Actually Complete? (E)	Year of Operation (F)
Program ABC	$1,000K	$900K	$1,000K	$950K	Year 1
	$2,000K	$2,100K	$2,000K	$1,950K	Year 2
	$3,250K	$3,600K	$3,250K	$3,300K	Year 3
	$4,750K	$5,600K	$4,750K	$4,700K	Year 4
	$5,250K	$6,600K	$5,250K	$5,300K	Year 5
Multi-Year Program Total	$16,250K	$18,800K	$16,250K	$16,200K	

graph is that Program B is spending more and achieving less than planned. This kind of plain insight is not easily available from other, less-objective reporting systems. What makes this even more useful is the fact that there is drill-down capability which allows the user to dive into the program's sub-projects to isolate where the underperformance stems from.

How to Create the Multi-Year Program Graph

This graph is constructed by plotting the cumulative data for the PV, EV, and AC of the program in each year. The data from every year of the program's execution is accumulated, as shown in Table 10.7.

Each of the curves is then plotted one by one. For example, the AC (see Column C in Table 10.7) was $900K in Year 1, $2,100K in Year 2, $3,600K in Year 3, and so on. Plotting these points cumulatively on the graph and connecting them gives us the AC curve seen in Figure 10.2.

Similarly, the PV (see Column D in Table 10.7) was $1,000K in Year 1, $2,000K in Year 2, $3,250K in Year 3, and so on. Plotting these points cumulatively on the graph and connecting them gives us the PV curve seen in Figure 10.2.

Finally, the EV (see Column E in Table 10.7) was $950K in Year 1, $1,950K in Year 2, $3,300K in Year 3, and so on. Plotting these points cumulatively on the graph and connecting them gives us the EV curve seen in Figure 10.2.

When these three curves are plotted, the visual representation of the program's multi-year journey is obtained, as seen in Figure 10.2. The above procedure needs to be done for every program that has a multi-year journey, in order to generate a graph for the program.

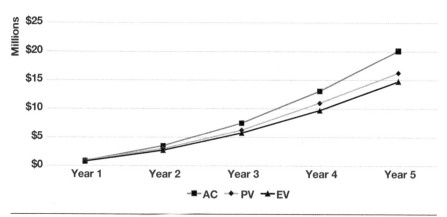

Figure 10.2   Multi-year performance graph of Program B

THE MULTI-YEAR PORTFOLIO VIEW

The multi-year portfolio view is a combination of all the aggregations seen in this chapter. It displays a view of the portfolio year by year, which can then be drilled down to a single year view of the portfolio showing the constituent programs and projects that were in the portfolio that year. The programs can then be drilled down further into individual projects. Table 10.8 shows the multi-year portfolio view with hyperlinks embedded in each of the years.

How to Create the Multi-Year Portfolio View

1. Each row in the multi-year portfolio view in Table 10.8 is obtained by summing up the projects in the portfolio for that year. For example, the totals row in Table 10.3 (constructed for Year 1) would become the Year 1 row for Table 10.8. Similarly, the totals row in Table 10.3 (constructed for Year 2) would become the Year 2 row for Table 10.8, and so on.
2. It's highly recommended to embed a hyperlink in the *year* entry in each row of Table 10.8. This hyperlink would point back to the single year portfolio view (Table 10.3) whose total is being used as a row in Table 10.8.
3. A portfolio's composition is bound to vary over the years. For example, in Year 1, it could have five projects, while in Year 2 it may have seven projects. This underscores why it's a good idea for the hyperlink to point to a detailed view that shows what projects were included in the portfolio for the year. This is also a good reason to keep accurate historical records.

The Multi-Year Portfolio View Rendered as a Graph

Although the multi-year portfolio view is useful in a table as shown in Table 10.8, it's more insightful to render the same information into a graph because that provides a more intuitive view, especially for portfolio governance (see later section in this chapter regarding presentation of aggregated data to portfolio governance). The multi-year portfolio graph is shown in Figure 10.3.

Since a portfolio has a multi-year journey in capability improvement, the graph artifact is a good fit for providing visibility into the portfolio's performance over the course of that journey. For example, one of the readily apparent insights from the graph shown in Figure 10.3 is that the portfolio is growing in size over the years, but slowly decreasing in its cost efficiency while showing an improvement in its schedule efficiency. This kind of vital insight is not easily available from other, less-objective reporting systems. What makes this even

Table 10.8 Multi-year view of the portfolio's performance

Name of the Portfolio (A)	Total Budget of the Portfolio (B)	How Much Money Did We Actually Spend? (C)	How Much Work Should Have Been Completed? (D)	How Much Work Did We Actually Complete? (E)	Year of Operation (F)	What is the Cost Efficiency? (G)	What is the Schedule Efficiency? (H)
	$10,000K	$9,000K	$10,000K	$8,000K	Year 1	1.06	0.95
	$20,000K	$21,000K	$20,000K	$18,500K	Year 2	0.93	0.98
Enterprise Portfolio	$32,500K	$36,000K	$32,500K	$30,000K	Year 3	0.92	1.02
	$47,500K	$56,000K	$47,500K	$42,000K	Year 4	0.84	0.99
	$52,500K	$66,000K	$52,500K	$53,000K	Year 5	0.80	1.01
Multi-year Portfolio Total	$162,500K	$188,000K	$162,500K	$151,500K		0.91	0.99

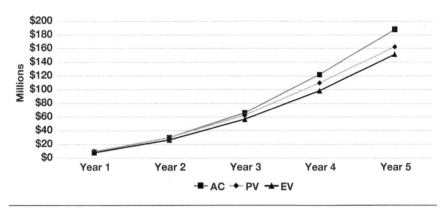

Figure 10.3   Multi-year performance graph of portfolio

more useful is the fact that there is drill-down capability that allows the user to dive into the portfolio's constituent projects to isolate the detailed performance in any given year.

How to Create the Multi-Year Portfolio Graph

This graph is constructed by plotting the cumulative data for the PV, EV, and AC of the portfolio in each year. The data from every year of the portfolio's execution is accumulated as shown in Table 10.8.

Each of the curves is then plotted one by one. For example, the AC (see Column C in Table 10.8) was $9,000K in Year 1, $21,000K in Year 2, $36,000K in Year 3, and so on. Plotting these points cumulatively on the graph and connecting them gives us the AC curve seen in Figure 10.3.

Similarly, the PV (see Column D in Table 10.8) was $10,000K in Year 1, $20,000K in Year 2, $32,500K in Year 3, and so on. Plotting these points cumulatively on the graph and connecting them gives us the PV curve seen in Figure 10.3.

Finally, the EV (see Column E in Table 10.8) was $9,500K in Year 1, $19,500K in Year 2, $33,000K in Year 3, and so on. Plotting these points cumulatively on the graph and connecting them gives us the EV curve seen in Figure 10.3.

When these three curves are plotted, the visual representation of the portfolio's multi-year journey is obtained as seen in Figure 10.3. If there is more than one portfolio (typical in large organizations), the aforementioned procedure needs to be done for every portfolio in order to generate a graph for the portfolio. This could then create a very useful picture that allows for the comparison between the different portfolios and their relative performance over the years.

CONSUMPTION OF AGGREGATED DATA BY PORTFOLIO GOVERNANCE

The main reason for aggregating data and creating these different views is to present the data to governance in a consumable form. This would then enable portfolio governance to make the right decisions and ensure that the portfolio meets its goals. We now explore the different scenarios as to how portfolio governance would consume the aggregated data.

Scenario 1: Review of Portfolio View by Portfolio Governance

In this scenario, the mEVM data of all the projects in the portfolio are shown to portfolio governance in a single table view, as first shown in Table 10.1. A slightly modified form is shown in Table 10.9 for convenience. At a glance, portfolio governance can obtain the following insights from looking at the table:

- Identification of projects that are spending as planned, as evidenced by high scores in cost efficiency (Column F)
- Identification of projects that are spending more (and achieving less) than planned, as evidenced by low scores in cost efficiency (Column F)
- Identification of projects that are executing on schedule, as evidenced by high scores in schedule efficiency (Column G)
- Identification of projects that are falling behind on schedule, as evidenced by low scores in schedule efficiency (Column G)

The information gleaned from Table 10.9 is crucially important because it filters the mass of all projects in the portfolio down to the few troubled projects that need attention and possible intervention by portfolio governance. By clicking on the hyperlink that is embedded in each project's name in Column A, portfolio governance can dive into the mEVM snapshot (see Figure 9.8 for an example) for that project for that month. The mEVM snapshot conveys a lot of useful information that enables portfolio governance to decide if the project is in serious trouble or is just going through a transient phase.

In this way, the portfolio view *boils down* the portfolio to the few projects that really need the time and attention of portfolio governance. This optimizes the use of portfolio governance's bandwidth and focuses their effort where it matters.

Table 10.9 Modified monthly portfolio dashboard showing year to date (YTD) values for AC, PV, and EV

Name of the Project (A)	Total Budget of the Project (B)	Actual Money Spent (YTD) (C)	Planned Work (YTD) (D)	Earned Value (YTD) (E)	Cost Efficiency (F)	Schedule Efficiency (G)
Project 1	$500K	$100K	$100K	$75K	0.75	0.75
Project 2	$750K	$300K	$200K	$300K	1	1.5
Project 3	$100K	$50K	$40K	$25K	0.5	0.625
Project 4	$240K	$200K	$175K	$175K	0.875	1
Project 5	$1,000K	$500K	$450K	$500K	1	1.11
...	...	...	...	...	...	...
...	...	...	...	...	...	...

Scenario 2: Review of Program View by Portfolio Governance

In this scenario, the mEVM data of all the projects in a particular program are shown to portfolio governance in a single table view, as was first shown in Table 10.2. A slightly modified form is shown in Table 10.10 for convenience. At a glance, portfolio governance can obtain the following insights from looking at the table:

- Summary readout of the program's overall cost performance (as evidenced by the program total cost efficiency score in the last row of Column F)
- Summary readout of the program's overall schedule performance (as evidenced by the program total schedule efficiency score in the last row of column G)

If the program as a whole is doing well on cost and schedule performance, portfolio governance may choose to stop there. However, if the program is not doing well on cost and/or schedule, portfolio governance may want to know which of the projects in the program are contributing to the underperformance. This can be done by reviewing the mEVM data of the constituent projects of the program:

- Identification of projects within the program that are spending more than planned, as evidenced by low scores in cost efficiency (Column F)
- Identification of projects within the program that are falling behind on schedule, as evidenced by low scores in schedule efficiency (Column G)
- In contrast, the projects within the program performing well on cost and schedule are evidenced by high scores in cost efficiency (Column F) and schedule efficiency (Column G), respectively

The program view is crucially important to portfolio governance in judging program performance, which is traditionally harder to gauge than project performance. However, the aforementioned view makes it possible to understand what's happening at the program level because it aggregates the performance of all of the projects in the program to a single number. Furthermore, this view enables portfolio governance to see at a glance which projects are contributing to the overall underperformance of the program.

By clicking on the hyperlink that is embedded in each project's name in Column A, portfolio governance can dive into the mEVM snapshot (see Figure 9.8 for an example) for an underperforming project and observe more closely why the project is underperforming and dragging down the program's numbers. It needs to be kept in mind that each program will need a program table.

Table 10.10 Modified monthly program view of Program ABC showing YTD values for AC, PV, and EV of the projects in the program

Name of the Project (A)	Total Budget of the Project (B)	Actual Money Spent (YTD) (C)	Planned Work (YTD) (D)	Earned Value (YTD) (E)	Cost Efficiency (F)	Schedule Efficiency (G)
Project 1	$100K	$50K	$40K	$25K	0.5	0.625
Project 2	$240K	$200K	$175K	$175K	0.875	1
Project 3	$1,000K	$500K	$450K	$500K	1	1.11
Project 4	$500K	$100K	$100K	$75K	0.75	0.75
Project 5	$750K	$300K	$200K	$300K	1	1.5
Program ABC's Total	$2,590K	$1,150K	$965K	$1,075K	0.93	1.11

As described before, the program view accomplishes two objectives—it provides visibility about the overall program's performance while preserving the option to understand which projects are underperforming and are in need of attention. As seen before, this optimizes the use of portfolio governance's bandwidth and focuses their effort where it matters.

Scenario 3: Review of the Multi-Year Project View by Portfolio Governance

In this scenario, the mEVM data for each of the years of operation for multi-year projects in the portfolio are shown to portfolio governance in a single table view, as first shown in Table 10.6. A slightly modified form is shown in Table 10.10 for convenience. At a glance, portfolio governance can obtain the following insights from looking at the table:

- Summary readout of the project's overall cost performance over all the years (as evidenced by the project's total score in the last row of Column G)
- Summary readout of the project's overall schedule performance over all the years (as evidenced by the project's total score in the last row of Column H)

The summary readout may be enough for portfolio governance to be assured that the long-running project is performing well and no further review is necessary. However, the following three sub-scenarios may drive a need for portfolio governance to know how the project performed in the different years:

- During the year, funding may be scarce and this multi-year project is one of the options to cut spend and divert to other projects
- As part of annual planning, all existing projects are compared against new demand to decide if this multi-year project should be continued for another year
- The multi-year project is not doing well on cost and schedule performance in the current year, spurring a discussion of whether this is a historical pattern (if the multi-year project has always been a bad performer, maybe now is the time to cut losses and terminate the project), or an atypical year for the project (some transient factors may be causing this historically sound project to underperform and these factors need to be rectified)

This can be done by reviewing the mEVM data of each of the years of operation of the project. Provided that all of the historical mEVM records are preserved, a hyperlink that is embedded into each of the years (Column F) should take the user to the mEVM data for that year for a more detailed look.

The multi-year project view accomplishes an important objective—it provides historical visibility about a long-running project, which enables a quick determination of the cost and schedule performance of this multi-year project. This determination provides a solid basis to decide if the multi-year project needs to be funded or terminated. As seen before, this rapid decision making optimizes the use of portfolio governance's bandwidth and focuses their effort where it matters.

Scenario 4: Review of the Multi-Year Portfolio View by Portfolio Governance

In this scenario, the performance of the entire portfolio over the years is shown to portfolio governance in a single table view, as first shown in Table 10.8. A slightly modified form is shown in Table 10.11, for convenience. This table provides objective evidence of how the enterprise portfolio has been performing over the years. At a glance, portfolio governance can obtain the following insights from looking at this table:

- An improvement in cost efficiency of the portfolio over the years is evidenced by increasing scores in Column F (this would mean that over the years, projects are performing better on costs)
- An improvement in schedule efficiency of the portfolio over the years is evidenced by increasing scores in Column G (this would mean that over the years, projects are performing better on schedule)
- A non-improvement or worsening trend in cost efficiency of the portfolio over the years is evidenced by unchanging or declining cost efficiency scores in Column F (this would mean that over the years, projects are not improving on cost performance)
- A non-improvement or worsening trend in schedule efficiency of the portfolio over the years is evidenced by unchanging or declining schedule efficiency scores in Column G (this would mean that over the years, projects are not improving on schedule performance)

Although the information in this table provides a high-level look, it offers a basic, data-driven answer to the fundamental question, "Is the enterprise portfolio improving over the years?" This answer is important to arrive at because of the huge amounts of time and money spent in improving the portfolio. The bulk of this book is focused on upgrading all of the portfolio building blocks from Level 1 to Level 2 or to Level 3. It is useful and appropriate to get a data-backed answer to whether all of those efforts are bearing fruit or not.

Table 10.11 Modified multi-year view of the enterprise portfolio for governance review

Name of the Portfolio (A)	Total Budget of Portfolio (B)	Actual Money Spent (C)	Planned Work (D)	Earned Value (E)	Cost Efficiency (F)	Schedule Efficiency (G)	Year of Operation (H)
	$10,000K	$9,000K	$10,000K	$9,500K	1.06	0.95	Year 1
	$20,000K	$21,000K	$20,000K	$19,500K	0.93	0.98	Year 2
Enterprise Portfolio	$32,500K	$36,000K	$32,500K	$33,000K	0.92	1.02	Year 3
	$47,500K	$56,000K	$47,500K	$47,000K	0.84	0.99	Year 4
	$52,500K	$66,000K	$52,500K	$53,000K	0.80	1.01	Year 5
Multi-Year Portfolio Total	**$16,250K**	**$18,800K**	**$16,250K**	**$16,200K**			

If the portfolio is indeed improving, this would be important to showcase to other executive stakeholders and organizational partners, such as Finance. Finance especially would be interested in having solid evidence that the portfolio is improving in cost and schedule efficiency over the years, which in essence shows that the organization's money is being spent in better ways.

If the data indicates that the portfolio is *not* improving, as evidenced by declining or flat scores in cost and schedule efficiency over the years, it constitutes an important wake-up call for portfolio governance that a close look needs to be taken at the current setup of the portfolio office. Professional portfolio consulting services may need to be engaged to obtain recommendations on improving the portfolio office and create a positive trend.

Success Factors in Portfolio Data Aggregation

As we saw, portfolio data yields very important insights when aggregated and presented to portfolio governance. The following factors are key to enabling portfolio data aggregation:

- **Factor #1—Commitment to mEVM**: mEVM is the foundation of all the aggregated views described in this chapter. Without the objective basis of mEVM, the whole principle of aggregation falls apart and cannot be used for any meaningful analysis. Any decision to use the aggregation methods described in this chapter would therefore need a strong commitment to implement mEVM reporting.

- **Factor #2—Universal application of mEVM**: Once the decision to implement mEVM is made, the portfolio office should ensure that all projects are reporting their status using mEVM. No exceptions can be made for any project or program because then it would not be possible to include that project or program in the aggregation, rendering the exercise useless.

- **Factor #3—Buy-in from portfolio governance**: The portfolio office needs to demonstrate to decision makers, including portfolio governance, the utility of aggregating portfolio data. This is best illustrated using the scenarios outlined before in the section titled *Consumption of Aggregated Data by Portfolio Governance*. Essentially, the decision makers have to be taught how to use the aggregated data. Once the potential and efficacy of aggregation is understood by portfolio governance, their support naturally follows and ensures success in the application of this technique.

- **Factor #4—Meticulous record keeping**: To be able to aggregate historical data going back several years, the data from those periods needs to be preserved. To ensure uniform preservation and availability, there is a need for the appropriate systems to capture and allow retrieval of data. SharePoint has been recommended extensively in this book as a tool of great simplicity and flexibility to accommodate these data retention needs. To ensure that data is not lost, there also needs to be disaster recovery policies in place that ensure the preservation of vital historical data. (For example: a backup copy of all the mEVM artifacts is made each month and preserved in an alternate location.)

- **Factor #5—Routine use of aggregated data**: For aggregation data to be used, it needs to be reviewed regularly. This, in return, requires a conscious decision on the part of the portfolio office to include the aggregated data in the portfolio materials that are given to the portfolio governance each month. When portfolio governance members start relying on the aggregated reports and interpreting the data contained in them, a reinforcing effect is created for the whole setup and ensures continuity of the practice.
- **Factor #6—Well-trained resources in the portfolio office**: It takes a fair amount of work to construct and maintain the aggregated portfolio views. Since projects are always executing and generating new mEVM data, this translates into a need to keep the view updated with the new information. There is also work involved in creating new hyperlinks and pointing them to the right places while ensuring that none of the legacy hyperlinks (such as the ones pointing to previous periods' mEVM details) are broken. It's important to ensure that all aggregation data is reviewed for correctness prior to distribution to portfolio governance.

LEVELS OF PORTFOLIO MATURITY

Level 1

- The organization has no concept of aggregation—data views are fragmented and limited to current year and current effort
- The need for aggregation may be felt, but no meaningful progress is possible because the prerequisites (for example, mEVM) are not in place
- Some aggregations are built and maintained with considerable effort (for example, projects under a program), but these aggregations have no foundational objective basis and cannot scale
- Portfolio governance is not acquainted with the concept of aggregation or its potential

Level 2

- The organization is becoming familiar with the concept of aggregation and creates aggregated data views in one or two dimensions of aggregation
- The need for aggregation has been felt and an objective method of reporting (such as mEVM) has been put in place, but the adoption of it may still be in progress and is not universal
- Portfolio governance is acquainted with the concept of aggregation and its potential, but all the desired artifacts to make it functional may not yet be developed

Level 3

- The organization is well acquainted with the concept of aggregation and creates aggregated data views that extend to all three dimensions of aggregation
- The need for aggregation has been felt and an objective method of reporting (such as mEVM) has been put in place, with universal and complete adoption
- Portfolio governance is familiar with the concept of aggregation and its potential
- Aggregated views are included regularly in portfolio reports and portfolio governance uses it to make decisions

CHAPTER SUMMARY

In this chapter, we began to build upon the foundation of mEVM in order to create impactful artifacts that produce truly useful insights to decision makers. The need for data aggregation was explored against the backdrop of the visibility-limiting nature of the annual funding cycle. The three main types of portfolio aggregation were listed before diving into a detailed treatment of how to create and maintain each type of view. There was also a detailed treatment that explores the scenarios of consumption of the portfolio data by portfolio governance and the intricacies thereof. Finally, the chapter concluded with the levels of portfolio capability.

Web
Added
Value™

This book has free material available for download from the
Web Added Value™ resource center at *www.jrosspub.com*

11

MEASURING STRATEGIC
ATTAINMENT USING mEVM

INTRODUCTION

In most large organizations, strategy occupies a significant mindshare among management, and a substantial amount of effort is spent in formulating and discussing strategy. In contrast, not much is done when it comes to measuring strategy attainment—partly because it is an abstract concept that does not easily lend itself to being measured. Strategy attainment deals with validating whether the planned strategy was implemented and at what cost. With modified earned value management (*mEVM*), it becomes possible to measure this previously abstract concept.

This chapter introduces an approach to objectively measure strategic attainment using mEVM and covers the following aspects:

1. Distinction between tactical and strategic attainment
2. Exploration of the nuances in measuring strategic attainment
3. Introduction of an approach to measure strategic attainment using mEVM
4. Exploration of a method to separate strategic activity from nonstrategic activity
5. Enumeration of the benefits of measuring strategic attainment using mEVM
6. Enumeration of the factors that promote success in measuring strategic attainment

TACTICAL ATTAINMENT VERSUS STRATEGIC ATTAINMENT

Before we approach the measurement of strategic attainment, we need to distinguish between tactical attainment and strategic attainment. For example, consider the difference between 10 projects delivering completed project charters that cost a combined $1M to produce versus one project delivering a usable mobile app that also costs $1M to implement.

While both efforts cost the same amount of real dollars, the charters are just intermediate by-products (with no intrinsic usable value), while the deployed mobile app has added strategic capability to the enterprise and, hence, has actual strategic value. From a strategic point of view, the deployed mobile app is significantly more valuable than the charters. Both artifacts have their own place in the value creation journey, but need to be valued differently.

THE COMPLICATED BUSINESS OF MEASURING STRATEGIC ATTAINMENT

Measuring strategic attainment is complex because there are several layers to the term. The concept of measuring strategic attainment could be modeled as consisting of two different main measurements—performing the implementation and fulfilling the objectives. Thus, the following questions should be answered:

- Was the strategy actually implemented? If the strategy was indeed implemented, were the cost and schedule parameters of the implementation under target, over target, or exactly as planned?
- Did the implemented strategy actually work? Did it achieve the declared strategic objectives?

Portfolio management can best answer the first question—whether the strategy was actually implemented and whether the implementation occurred on time and on budget. The second question can only be answered by measuring the key performance indicators that were identified as being impacted by the strategy. It may also be possible to answer the second question using the portfolio benefits management process as outlined in Chapter 7. This chapter focuses only on the first question, namely: "Was the planned strategy actually implemented and if so, what were the cost and time parameters of implementation."

Measuring Strategy Attainment Using mEVM

To illustrate the concept, we'll need to use a business case involving a fictional company as described here: Safe and Sound, Inc. is an insurance company that has been around for many decades. They have a loyal user base and their modest information systems can be classified as legacy. All their transactions are mail based—payments, insurance IDs, claims, and payments are all sent through the mail. With new management taking over the helm, the company hires a strategy consulting firm to chart a new direction.

The strategy consulting firm recommends implementing the following strategy road map to modernize and achieve parity with competitors. The journey to modernization includes making several *strategic hops* over a multi-year period. Each of these *hops* is represented by a strategic priority as outlined here:

- **Strategic Priority 1**: Real-time payment: Enable users to pay premiums online (Year 1)
- **Strategic Priority 2**: Self-service for documentation: Enable users to print ID cards and proof of insurance through an online portal (Year 2)
- **Strategic Priority 3**: Self-service for claims: Enable users to submit claims online, including documentation (Year 3)
- **Strategic Priority 4**: Multi-platform capability: Enable users to do all of the above on a mobile platform in addition to online (Year 4)

Starting with Strategic Priority 1 and decomposing that priority into discrete projects would yield the information shown in Table 11.1.

The following is an explanation of the structure of the table:

1. The strategic priority was decomposed into three sub-projects (this is an example only—in reality, it could be more or fewer, depending on the particular priority). Each of these sub-projects (hereafter referred to as projects) delivers a concrete strategic capability without which the overall strategic priority cannot be achieved.
2. A round number of $1M was chosen as the budget for each project, for ease of use.
3. For simplicity, let's assume the three projects run concurrently with a planned start date of January 1 and a planned finish date of December 31 of Year 1.
4. The planned value (PV) of each project is equal to the project's budget.

Continuing the aforementioned process for all of the strategic priorities would yield the complete strategic decomposition table as shown in Table 11.2. Note that for the strategic priorities numbered 2, 3, and 4, the sub-projects are not named and are just depicted by a numerical suffix.

Representing the data of Table 11.2 in graphical format, we see the graph shown in Figure 11.1.

Table 11.1 Partial strategic priority decomposition table

Strategic Priority	Strategic Priority Description	Sub-Projects	Sub-Project Budget	Planned Value	Estimated Completion
1	Real-time payment processing: enable users to pay premiums online	Revamp website to accept online payment	$1M	$1M	Year 1
		Implement payment gateway interface	$1M	$1M	Year 1
		Update customer database schema to allow for online access	$1M	$1M	Year 1

Table 11.2 Complete strategic priority decomposition table

Strategic Priority	Description	Sub-Projects	Sub-Project Budget	Planned Value	Estimated Completion
1	Real-time payment processing: enable users to pay premiums online	Revamp website to accept online payment (Sub-project 1)	$1M	$1M	Year 1
		Implement payment gateway interface (Sub-project 2)	$1M	$1M	Year 1
		Update customer database schema to allow for online access (Sub-project 3)	$1M	$1M	Year 1
2	Self-service for documentation: enable user to print ID cards and proof of insurance through an online portal	Sub-project 4	$1M	$1M	Year 2
		Sub-project 5	$1M	$1M	Year 2
		Sub-project 6	$1M	$1M	Year 2
3	Self-service for claims: enable users to submit claims online including documentation	Sub-project 7	$1M	$1M	Year 3
		Sub-project 8	$1M	$1M	Year 3
		Sub-project 9	$1M	$1M	Year 3
4	Multi-platform capability: enable users to do all of the above on a mobile platform in addition to online	Sub-project 10	$1M	$1M	Year 4
		Sub-project 11	$1M	$1M	Year 4
		Sub-project 12	$1M	$1M	Year 4

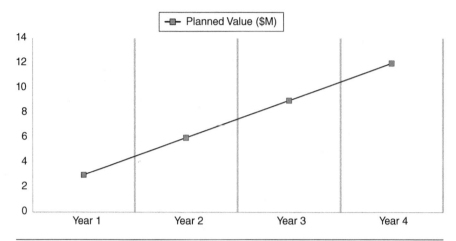

Figure 11.1   Graph showing PV of strategic projects across multiple years

Some key points to note about the graph are outlined here:

1. The X axis shows the multi-year journey of the strategic road map
2. The Y axis shows the PV of the strategic road map
3. The PV curve is cumulatively plotted year over year (For example: Year 1 has $3M of PV, Year 2 has $3M on top of that, making it $6M, and so on.)

Now, let's overlay on this curve the actual dollars spent by the projects. The actual dollars spent are shown in Table 11.3. When these values are plotted in the graph, the resulting graph is shown in Figure 11.2.

A quick study of the graph indicates that the actual cost (AC) is running a little high as compared to the PV. But the true value of the graph lies in plotting the strategic earned value (SEV) on top of these two curves. We'll first need to define what SEV is:

1. Recall the mEVM rules of recognizing earned value (EV) from Chapter 9. As a recap, we only recognize the full (100%) EV for a deliverable when the deliverable is 100% complete.
2. SEV is very similar, with the difference being that it is recognized only when the strategic capability is 100% delivered to the enterprise. For example, the SEV of the project *Implement payment gateway interface (Sub-project 2)* is only realized when the payment gateway is implemented and functional.
3. If most of the project is done but the end result is that the payment gateway is not available to accept payments, then the strategic purpose

Table 11.3   Yearly spend of each sub-project in the strategic priority decomposition table

Project	Actual Dollars Spent (in $M)	Year of Spend
Sub-project 1	$1M	Year 1
Sub-project 2	$1.5M	Year 1
Sub-project 3	$1.5M	Year 1
Sub-project 4	$2M	Year 2
Sub-project 5	$1M	Year 2
Sub-project 6	$1.5M	Year 2
Sub-project 7	$1.5M	Year 3
Sub-project 8	$1M	Year 3
Sub-project 9	$2M	Year 3
Sub-project 10	$2M	Year 4
Sub-project 11	$1.5M	Year 4
Sub-project 12	$1.5M	Year 4

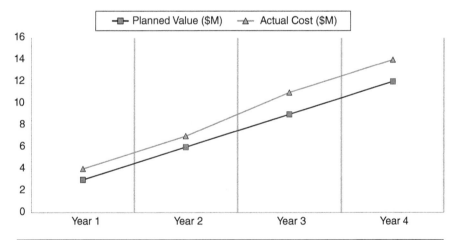

Figure 11.2   Graph showing PV and AC of strategic projects across multiple years

is defeated and the SEV cannot be recognized (i.e., it would be zero). Unlike regular EV, SEV cannot be partially granted for partially finished milestones.

4. Note that the maximum SEV for any project is equal to the PV, which is equal to the budget.

Consider Table 11.4, which shows the actual strategic value attained by the 12 projects that make up the four strategic priorities. The column showing the actual dollars spent in each sub-project is taken from Table 11.3 and included in Table 11.4. The zeros against sub-projects 5, 6, 9, 11, and 12 represent situations where the SEV could not be attained/recognized because the strategic capability was not delivered in a usable fashion to the enterprise.

Finally, the SEV data is plotted on the same graph that shows PV and AC—and the result is shown in Figure 11.3. Some conclusions from Figure 11.3 are listed here:

- **Takeaway #1**: SEV is the least of the curves while AC is the highest of all the curves—this means that we're spending more money than planned on strategic projects and achieving less than planned.
- **Takeaway #2**: The reduced SEV can be attributed to strategic projects not actually delivering the strategic objective to the enterprise. While the reality may be more complex than previously described, the fact remains that the strategic initiatives are not delivering as promised.
- **Takeaway #3**: The increased costs are indicative of ineffective project spend management. This is evidenced by the AC curve being highest of all.

Table 11.4 SEV attainment for each sub-project along with year of spend

Project	Actual Dollars Spent (in $M)	Year of Spend	Strategic Earned Value Attained (Upon Full Completion)
Sub-project 1	$1M	Year 1	$1M
Sub-project 2	$1.5M	Year 1	$1M
Sub-project 3	$1.5M	Year 1	$1M
Sub-project 4	$2M	Year 2	$1M
Sub-project 5	$1M	Year 2	0
Sub-project 6	$1.5M	Year 2	0
Sub-project 7	$1.5M	Year 3	$1M
Sub-project 8	$1M	Year 3	$1M
Sub-project 9	$2M	Year 3	0
Sub-project 10	$2M	Year 4	$1M
Sub-project 11	$1.5M	Year 4	0
Sub-project 12	$1.5M	Year 4	0

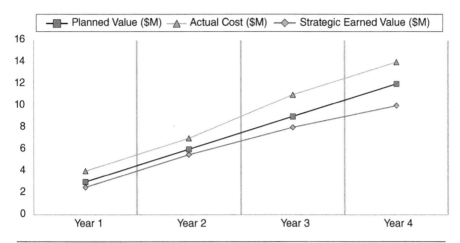

Figure 11.3 Graph showing PV, AC, and SEV of strategic projects across multiple years

HOW MUCH OF OUR ACTIVITY IS STRATEGIC?

Every leader wants to know if their organization is working *on the things that matter*. While a good mix of tactical versus strategic is recommended, this mix is only possible to achieve after knowing the current actual ratio of tactical to strategic. Here is an interesting view that is created using the SEV concept.

The graph shown in Figure 11.3 was created by overlaying the actual spend of only the sub-projects that were identified in the original strategic table (Table 11.2). What would happen if we instead took *all* of the project spend and rendered that in Figure 11.3? However, we would keep the PV and SEV unchanged. (In other words, showing only the PV and SEV of strategic projects but the AC of all projects.) In all probability, it would look like the graph shown in Figure 11.4.

Conclusions from the data shown in Figure 11.4 are as follows:

1. The combined AC of *all* projects is far above the PV of the *strategic* projects.
2. The combined AC of *all* projects is far above the SEV of the *strategic* projects.

From these conclusions, some or all of the following factors seem to be at play in the organization:

1. **Factor 1**: The organization is spending a lot of money on projects that are not strategic; unless this is a planned objective (to emphasize tactical priorities over strategic), this is a bad situation

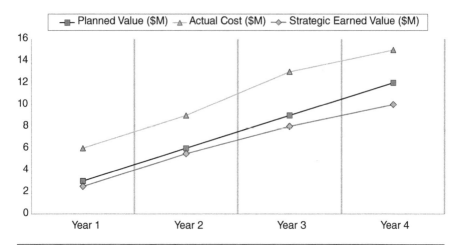

Figure 11.4   Graph showing PV and SEV of strategic projects against total cost of all projects across multiple years

2. **Factor 2**: Strategic vision is not well articulated—strategic planning seems to be an abstract exercise without a validated connection to execution
3. **Factor 3**: The strategic road map is not decomposed completely into projects—there are gaps that result in strategic projects not being identified properly
4. **Factor 4**: Some strategic projects are not delivering the previously declared usable strategic capability to the organization
5. **Factor 5**: The projects seem to be coming in over budget as a pattern

Identification of these factors is a huge benefit to a chief information officer (CIO) or other executive management, because they are early warning indicators that the strategic agenda is likely headed for failure or sub-optimal achievement. In my experience, this data is rarely made available to the CIO in a time frame when the trajectory can be altered.

BENEFITS OF MEASURING STRATEGIC ATTAINMENT USING mEVM

Benefit #1: Detects Possible Inadequacy of Strategic Plan

The whole technique of strategic attainment measurement relies on the existence of a sound, validated, modular strategic plan. While deploying this technique, it may become apparent that there is simply not a sound strategy to build anything upon. There may be a plan, but it may not be a realistic and actionable plan to measure progress against. If the strategic plan itself is unrealistic or insufficiently articulated, the probability of successfully executing the plan is low. Pointing out the deficiency in the strategic plan may be a painful but useful benefit of deploying this system of measurement.

Benefit #2: Provides Concrete Definition of the Strategic Road Map in Terms of Discrete Projects

Few organizations have a clear view of how their individual projects add up to form the strategic road map. Even if this is known to a few people, the rank and file have no conceptualization of the big picture. This fragmented understanding results in lack of understanding about what is strategic and what is not. This feeds the general idea that strategic planning is not a useful activity and no one wants to waste any time on it. With this technique, a clear and simple connection is established between projects and strategy.

Benefit #3: Enables Revalidation of Strategic Investment during Annual Planning

Once the link is established between projects and strategy, it is easy to show the total strategic demand during annual planning (this is done by adding up the annual planning demand of the sub-projects that appear in the strategy table). It is then possible to show how much of this demand is finally funded during annual planning. This brings into focus a clear question: "Are we funding the strategic demand in full for the new year? If not, why not?" Visibility to this choice enables the right allocation of funds during annual planning.

Benefit #4: Draws Distinction between Tactical Value and Strategic Value

As explained earlier in this chapter, there is a clear difference between tactical and strategic value. As a result of this strategy attainment measurement, the organization is able to identify the two different kinds of value and ensure that strategic value is realized at all times. By identifying this difference, the organization can answer the important question: "How much of our work is strategic?" The ratio of EV to SEV is a good proxy for measuring the ratio of tactical value to strategic value.

Benefit #5: Provides Useful Comparison between Strategic Demand and Strategic Supply

This technique shows what it truly costs (the demand) to produce strategic transformation (provided the strategic table is comprehensively developed, cross-checked, and validated). It then goes on to show how much the organization decided to invest in real dollars (the supply) toward achieving this strategic transformation. Why is this important? In the maze of individual projects, programs, and their funding decisions, this critical visibility of strategic supply versus strategic demand is often lost. This technique brings back that visibility.

Benefit #6: Provides Useful Comparison between Tactical Demand and Strategic Demand

Every organization tries to strike the right balance between the tactical and the strategic. However, few organizations actually have the data that enables them to make an objective comparison and conscious choice. With strategy attainment measurement, a distinction is drawn between what is strategic and what is tactical (any project that does not appear on the strategy decomposition table is likely tactical). The tactical demand is then compared with the strategic demand, allowing executive management to decide where they want to invest and how much. This feedback loop could also spark a re-examination of the strategic decomposition table to identify gaps (both false negatives and false positives) creating a higher quality strategy artifact that reflects reality.

Benefit #7: Provides Useful Comparison between Strategic PV and SEV

This comparison could be one of the most valuable for executive management— it answers the question: "How are we really doing on our strategic journey?" Historically, this has always been a difficult question for organizations to answer. The reason lies in the mismatch between strategy (often an abstract, aspirational concept) and projects (concrete efforts with real deliverables). With this technique, a robust crosswalk is created to bridge strategy to projects, and then the SEV validates whether the strategic capability delivery actually happened.

Benefit #8: Provides Useful Comparison between the AC of Strategic Projects and the Actual Attainment of SEV

Another informative comparison for management is to see what the true cost of implementing the strategy is. This visibility is obscured in the overflows and underflows of various projects, thus removing a useful insight about the cost of the strategic transformation. This technique provides that visibility and compares the actual cost incurred by strategic projects to the attainment of SEV.

Benefit #9: Preserves a Multi-Year View of the Progress Made against a Road Map

One of the reasons why strategy attainment is hard to measure is that the annual financial cycle of funding projects breaks up the continuity of tracking. With the technique of SEV and the tracking mechanism, the visibility is preserved and enables executive management to see the longer-term trend in strategy execution. This technique also enables decision makers at annual planning to see past performance and use that as an input for funding decisions. Finally, the multi-year view enables the portfolio office to demonstrate the value of intervention in correcting the trajectory of underperformance.

FACTORS OF SUCCESS IN MEASURING STRATEGIC ATTAINMENT

While strategy attainment measurement is a very powerful technique, there are some best practices to ensure that it delivers the promised benefits in terms of visibility and control. The following are some of the proven predictors of success in implementing this technique.

Factor #1: Discipline in Decomposing the Strategy to Projects

The foundation of the whole process is to decompose the strategy to discrete, actual projects with scope, schedule, and budget. This needs to be done in a comprehensive way, avoiding the two common problems—false negatives (leaving strategic projects off the list) and false positives (including nonstrategic projects on the list).

Factor #2: Consensus about the Strategy Table

The strategy decomposition table needs to be shared with all stakeholders and everyone needs to agree that the strategy table is accurate. While building consensus is important, care must be taken to avoid the insertion of nonstrategic projects by some stakeholders who perceive a prioritizing advantage to classifying their pet projects as *strategic*. At the same time, care must be taken to avoid leaving off strategic projects.

Factor #3: Precision in Identifying *What* Will Be Delivered Strategically

A common challenge in most organizations is that people are unable to get specific in identifying what the strategic deliverable is. For example, consider a strategic goal that reads like: *synergize multi-channel availability to one ubiquitous interface*—a word soup that could mean almost anything and is very hard to measure. Consider the more direct specification of the same goal: *allow customers to submit their information through mail, internet, or mobile device using the same form*—this is straightforward to measure. For example, the status report could be: *mail and internet availability accomplished; mobile not yet delivered.*

Projects that are unable to specify their strategic offering may actually be providing a clue that the project may not be worth doing. Executive management needs to establish a clear expectation with project owners that their project's strategic offering needs to be specific, straightforward, and measurable in delivered/not delivered terms. Projects that cannot offer up a strategic deliverable should not be on the strategy table.

Factor #4: Firm Guidelines for Approving SEV

Projects that do not deliver the promised strategic capability should not be allowed to claim the SEV. This could become a politically charged topic with project owners pointing the blame at others for not being able to deliver the promised value. It could also lead to political jockeying with the aim of obtaining SEV for projects that *almost delivered.*

Therefore, executive management needs to specify very clearly that the only way of obtaining SEV is to deliver a usable strategic capability in production. If there are valid reasons for something not making its way to production, that should be investigated and followed up, but no exceptions should be made in granting SEV. This is the only way to show the true state of affairs about progress made on the strategic road map.

LEVELS OF MATURITY

Level 1

Although there is substantial mention of strategy, the concept of tracking strategic value attainment may be absent in the organization:

- There is no distinction between strategic and tactical attainment
- mEVM may not be rolled out in the organization or may be in its infancy
- The concept of SEV may be completely unknown

- Due to all of the aforementioned reasons, there may not be an objective way to classify the project activity in the organization as strategic versus tactical

Level 2

- An awareness of the concept of tracking strategic value attainment exists, but it is not a well-grounded concept—hence, it is not done uniformly or comprehensively
- There may be an awareness of the distinction between strategic and tactical attainment; however, this distinction may be in its infancy and hence, unable to be standardized or reported on
- mEVM may have been rolled out in the organization, but not yet deployed to measure strategic attainment
- The concept of SEV may be known, but not objectively defined in a way to ensure tracking
- Due to all the still-forming capabilities listed here, the organization is not yet able to meaningfully track strategic attainment in a uniform manner

Level 3

- The concept of tracking strategic value attainment is well known and is implemented uniformly and comprehensively across the organization
- The distinction between strategic and tactical attainment is well known and reflected in artifacts
- mEVM is rolled out to the organization and mEVM artifacts are available for deployment to measure strategic attainment
- The concept of SEV is well known and objectively defined; artifacts are designed to capture and report on SEV, which is regularly disseminated to executives
- Due to all of the previously mentioned capabilities being in place, the organization is able to meaningfully track strategic attainment
- Executive leadership is receptive to the insights provided by SEV analysis and are willing to take actions to ensure an upward trend in improving strategic attainment

CHAPTER SUMMARY

In this chapter, we began to explore the difference between tactical value and strategic attainment. We proceeded to define strategic attainment and the different ways to measure that dimension. We then revisited the strategic decomposition table and introduced an approach to measure strategic attainment using mEVM. Using this approach, we proposed a method to separate strategic activity from nonstrategic activity—a view that is bound to interest executive decision makers. By rendering all of the data into a graph, we obtained a visual representation of SEV and proceeded to list the benefits of tracking the same. We covered the factors that spell success in tracking SEV and, finally, rounded out the chapter with a look at portfolio levels of maturity.

This book has free material available for download from the
Web Added Value™ resource center at *www.jrosspub.com*

12

HOW TO ROLL OUT mEVM TO THE ENTERPRISE

INTRODUCTION

In the last two chapters, the concept of modified earned value management (*mEVM*) was introduced and the artifacts of mEVM were detailed at length. In this chapter, the process of how to exactly roll out mEVM in an organization will be discussed. As part of that discussion, the following topics are covered:

1. Detailing the tactical list of steps needed to deploy mEVM in a portfolio setting
2. Short-, medium-, and long-term considerations to keep in mind during the rollout
3. Discussion of basic and advanced evasive tactics that are typically seen in project stakeholders who have been asked to adopt mEVM
4. Introduction to the enhanced monitoring list (EML)
5. We also cover the typical journey of a portfolio that has started using mEVM and what the portfolio manager can expect to see

STEPS TO ROLLING OUT mEVM REPORTING FOR THE PORTFOLIO

The primary step to take before attempting to roll out mEVM is to get solid buy-in from leadership. It's the foundation upon which a complete rollout is achieved. In order to get buy-in from leadership, the following steps need to be taken:

- Explanation of mEVM as a proven, decades-old technique to spot objective performance

- A clear statement of how mEVM could help the portfolio overcome its biggest pain point (every portfolio has this problem of not knowing how its projects are truly performing and therefore have difficulty in keep/kill decisions)
- Assurance that the mEVM implementation will be as *light-footprint* as possible

Step 1: Holding an Introductory Workshop on mEVM

The first step in rolling out mEVM is to hold a comprehensive workshop introducing it to the project managers. The content from Chapters 8 and 9 of this book can be readily repurposed to create the training material. Some skepticism can be expected from the stakeholders—mainly centered on using a new technique for measuring project performance. Once again, it's important to secure management support for this rollout and fend off some inevitable mumbling/grumbling that results when people are asked to change their method of functioning. During the workshop, the portfolio office needs to make sure to provide the audience with a view of the mEVM implementation rollout that is ahead (discussed in Steps 2 through 6 of the upcoming text). It's also important to emphasize to the project manager community the benefits that would come their way by adopting mEVM—namely, better predictability and control over their projects.

Step 2: Creation of the mEVM Templates

Next, we need to capture each project in the mEVM's Excel template that was discussed previously in Chapter 9. This activity has to be done by the project manager, but in close association with the portfolio manager. This will take a few iterations, as all of the *finer points* of the template will have to be considered (see Chapter 9 for a list of things to observe while creating the Excel template). The mEVM template will need to be created for all of the projects in the portfolio. However, this is a one-time activity.

Step 3: Creation of the mEVM Graphs

Creating the mEVM graphs would be the next step. The data for the graph for a project would come from the Excel template. From the graphs, the current reading of cost variance (CV) and schedule variance (SV) is calculated (the graphs are created once, but will need to be updated every month).

Step 4: Creation of the mEVM Reporting Package

Create the mEVM snapshot slide for the month. As explained in Chapter 9, this is a concise slide that contains the following:

- The graph as it stands for the month
- A readout of the SV and CV for the project up to that point
- A brief explanation of why the SV and CV are showing what they're showing—essentially, an explanation of the variance and a bullet or two about what the *get-well* plan is (see Figure 9.10 for reference)

Step 5: Creation of the mEVM Portfolio View Table

Create the reporting table with all the projects and their corresponding SV and CV. It would also help to create the hyperlinks that connect each project on the table to its detailed mEVM slide. The reporting table should be uploaded to SharePoint (see Table 10.1 in Chapter 10—also note the detailed procedure to create this table in that chapter).

Step 6: Holding the mEVM Workshop for Governance

Hold an mEVM workshop (with the live data) for the portfolio governance council and other decision makers including senior management. As part of that workshop, include the following content:

- A refresher to mEVM—can use a concise version of the training material shown to the project managers
- How to read the mEVM data—acquaint the audience with how to interpret the data being shown (refer to Chapters 8 and 9 regarding how to read the mEVM graphs)
- Explain to decision makers what questions need to be asked from the data being shown
- Explain how projects that show a worsening trend will be moved to the EML

IMPORTANT CONSIDERATIONS TO KEEP IN MIND DURING THE ROLLOUT

Short-Term Considerations

Although the six steps laid out in the previous section seem straightforward, expect the audience to take some time understanding how it all works. Patience is key in waiting for the organization to adopt the system. If your organization has a change management function, it's recommended to use them in rolling this out. Here are some of the short-term realities a portfolio manager has to deal with during the initial days of the mEVM rollout:

- Expect to do a lot of handholding. For example, in Step 2, the project manager is expected to create the mEVM template from their project's deliverables. However, in reality, the portfolio manager will have to create at least the first draft based on the project plan. Then, a few iterative meetings later, a final version of the mEVM template emerges for that project.
- The mEVM template is the cornerstone of mEVM reporting. It's important to socialize the mEVM template of each project with all the stakeholders of that project and get their signoff.
- Every project may not get their mEVM artifacts ready to start reporting at the governance meeting. You may have to start reporting on some projects and wait for the others to catch up. In fact, this may be a good thing—it has been noticed that people take their mEVM artifact creation seriously once they start seeing how the artifacts are being reported on (for other projects).
- The portfolio manager also needs to spend time with the portfolio governance members to help them understand the data that mEVM is producing. Just like the project teams take time to *get mEVM*, the governance members also need some time and guidance to understand how to react to the data.
- The portfolio governance/decision makers need to be quite *forgiving* during the initial days of mEVM rollout. The reason, as explained here, is twofold. First, the system needs to take root in the organization. Second, projects start doing better with mEVM in a matter of a few months. It's prudent to wait for the mEVM effect before taking punitive measures for underperformers.
- You can also expect to see projects claiming earned value (EV) without completing the deliverable in question. Bad habits from the past may manifest, as projects deal with the reality of mEVM (see an upcoming section relating to *common evasive tactics related to mEVM and countermeasures*).

Medium-Term Considerations

The *short-term phase* of mEVM lasts anywhere from six to nine months, depending on the organization. We then move to the medium-term phase of the mEVM implementation.

- This phase is characterized by regular mEVM readouts for each project and the project managers becoming familiar with mEVM terms and artifacts. At this point, the portfolio manager/governance needs to start becoming rigorous with mEVM. This means focusing on underperforming projects, looking at the trend line, and asking hard questions such as:
  - Is the CV or SV getting worse with each readout?
  - Is the *get well* plan of the project working or are things getting worse?
  - If the project seems to be ahead in cost and schedule, should they be returning some funds back to the portfolio? Etc.
- Now would also be a good time to start introducing the EML—also known as the *kill list*. This provides projects the necessary incentive to course correct if at all possible.
- For distressed projects, now may be the right time to do a reset. In essence, this is a fresh start for the project. It may involve recreating the mEVM template, possibly ignoring or writing off the previous sunk cost, redoing the mEVM graph, and starting the reporting fresh.
- One can also see the beneficial effects of mEVM begin to appear. The well-run projects will take to mEVM easily and their mEVM data will reflect their good performance. With the help of mEVM data, project and program managers (of well-running projects) begin to realize the true impediments to their projects' critical paths, and take action accordingly.

Long-Term Considerations

The long-term effects of mEVM are to produce a high performance portfolio with all of the players understanding the need for projects to run optimally and to take swift action to remediate whenever this is not the case.

- Projects produce high quality mEVM templates that are deliverable focused. The portfolio manager has to do little, if any, follow up to ensure the templates meet the standard.
- There is a general understanding/expectation among the project manager community that their projects' performance will be closely tracked and that they should be ready to explain their variances as reported by mEVM.
- The enhanced visibility created by mEVM creates a political climate of accountability, especially if there also exists a strong governance aspect to the portfolio that is not afraid to act on the conclusions of mEVM data.

- Advanced uses of mEVM data are made, including tie-ins with strategic attainment (see Chapter 11).

In the long term, mEVM creates a nimble and efficient portfolio with optimal allocation of resources.

COMMON EVASIVE TACTICS RELATED TO mEVM AND COUNTERMEASURES

An interesting dynamic develops when mEVM is implemented in a portfolio. Things that were previously hidden are brought into focus, and the project teams adopt various tactics in response to this situation. These tactics are divided into two sections—basic evasive behaviors seen during the initial phases of mEVM rollout and the advanced evasive behaviors seen as mEVM gets well entrenched. To be fair, some of the situations described in the following text are not *tactics* per se—they are simply situations that need to be solved, preferably in a partnership between the portfolio team and the project teams.

Basic Evasion Tactics

Tactic #1: I Have No Time to Do mEVM

This is a popular tactic, especially as the portfolio office tries to roll out mEVM. "My project is so critically important and we are so far behind plan that we cannot afford to do mEVM"—thus goes the reasoning. While bordering on the absurd, this is a tactic that will have to be dealt with in order to move forward with the portfolio capability improvement. Countermeasures include the following:

- Portfolio governance needs to get buy-in from all levels of management that is a *must-do*. The portfolio then needs to lay down the management expectation that all projects need to follow mEVM—funding may be made contingent on adopting mEVM (a strong incentive that cannot be ignored).
- Funding may need to be a lever that drives adoption. In other words, projects must adopt mEVM as a prerequisite to approach the portfolio for funding. Naturally, this would not be feasible in a funding model where the portfolio is not responsible for disbursing funds.
- Finally, one effective countermeasure has been to *name and shame*. A list of all projects are shown in the portfolio meetings—and then, the project owners who do not have their mEVM in place are asked to explain

themselves. Surprisingly, the politics of parity exert a compelling pressure to get the project owners to do mEVM.

- In summary, this is a tactic that will have to be countered in order to make any progress. To overcome this roadblock, the portfolio team could go the extra mile in helping the project create the mEVM artifacts. As an ultimate measure, executive intervention may be necessary.

Tactic #2: I Don't Need to Do mEVM, I Already Have a Comprehensive Project Plan

This is another classic evasion tactic—meant to preserve the status quo by holding up a false equivalent to mEVM. The reasoning is somewhat along the lines of: "We're already running well, so we don't need enhanced or remedial measures such as mEVM." The problem is that no one knows how the project is actually doing without an objective system of measurement such as mEVM. And a *project plan* could mean a wide range of techniques, with some being quite ineffective. So the basic premise of *we're doing well* may not even be valid. In any case, a situation where some people are doing mEVM and some not, is a recipe for confusion in a portfolio. It simply won't scale and cannot provide portfolio level inputs that are important to the portfolio decision makers. Countermeasures include the following:

- As previously mentioned, any rollout of mEVM needs to be preceded by a comprehensive agreement at all levels of management that this is a must-do. The portfolio then needs to lay down the management expectation that all projects need to follow mEVM—funding may be made contingent on adopting mEVM (a strong incentive that cannot be ignored).
- Again, funding may need to be a lever that drives adoption—as in, projects must adopt mEVM as a prerequisite to approach the portfolio for funding.
- One response that has been found to be quite effective is to take the *comprehensive project plan* and create mEVM artifacts out of it. It would involve the portfolio team taking the lead in creating the mEVM artifacts and getting the project team to sign off on the same. The project team will hardly have any grounds to object since they are being taken at their word ("We have a comprehensive project plan") and the same is being used to create an mEVM equivalent.
- Typically, such a plan would start underperforming in a couple of cycles, forcing the project team to do a proper mEVM plan. This would also spotlight any projects that are candidates for the EML or the kill list.

Tactic #3: Happy to Do mEVM, but Cannot Provide an mEVM Plan Yet

One of the typical responses encountered during mEVM rollout is: "I'll do mEVM later; right now I need to do the project." This is almost like saying, "I need to work without a plan before I can make a plan." While it's true that there may be urgent things that need to be done, postponing the mEVM plan is invariably a bad idea. First, the basic mEVM template is typically a one-pager and can be created in very short order. Second, the mEVM template can be modified as necessary. So there is no real reason to postpone the mEVM plan. Countermeasures include:

- As seen in previous cases, portfolio governance needs to make their expectation clear that every project needs to have their mEVM ready.
- The portfolio team can take the lead in creating an *initial* mEVM plan and getting the project team to sign off on the same. Having an *initial* mEVM plan in place will accelerate the project in creating a *permanent* mEVM plan that reflects a more accurate understanding of the project's scope.

Tactic #4: Using Vague Milestones

After the project teams understand that there is no way of avoiding mEVM, the evasion becomes a little more subtle. The mEVM technique is adopted, but the milestones chosen are vague enough to be interpreted in different ways. It's also quite possible that some teams are not actively trying to circumvent the system—it's how they have always done things and part of the reason why so many projects are in trouble. What do *vague milestones* mean? For example, consider the difference between these two milestones:

- Milestone #1: *Strategize approach toward new market offering*
- Milestone #2: *Publish document describing strategy toward penetrating new market for product XYX*

Milestone #1 is quite vague compared to Milestone #2, in the following ways:

- It's not clear what will be created/delivered
- It's hard to say with certainty whether this activity was completed or not—simply not a binary task whose completion can be categorized as yes/no.

Imagine several vague milestones such as the ones mentioned here in an mEVM template—the whole point of mEVM would be thwarted because it would be very hard to declare these milestones as done or complete. Countermeasures include:

- At least in the beginning stages of mEVM rollout, the portfolio manager will need to carefully screen each mEVM template to spot vague milestones. Many project managers will need some kind of mentoring to spot and recast vague milestones in terms of concrete deliverables with binary completion status.
- After vague milestones are fixed, the mEVM template needs to be approved by project owners to ensure that they stand behind the new milestones.
- The problem of vague milestones typically goes away once project managers begin to understand mEVM.

Tactic #5: Using Milestones of Activity Instead of Attainment

Using milestones of activity is a common problem found when people try to create mEVM templates for their projects. This is somewhat related to the mindset that creates the vague milestones covered in the previous section. What does this mean exactly? Consider the following:

- Milestone #1: *Meet with stakeholders to obtain direction on new market strategy*
- Milestone #2: *Acquire stakeholder signoff on document concerning the direction of new market strategy*

The difference between the two should be apparent:

- The first milestone, while not vague, is hard to pin down in terms of attaining the original purpose. For example, the *meet with stakeholders* could occur and not result in everyone signing off on the direction; in which case that work will still need to be done at additional effort and cost.
- The second milestone is very specific, with no room for misconstruction—either the stakeholder signoff was obtained or it wasn't.

Having milestones of activity is a prominent cause for obscuring the real state of the project. While we could be *checking off the boxes* in terms of completing various activities, the corresponding deliverables may not be complete, setting the project up for future trouble. The countermeasures for this problem are the same as the ones for vague milestones:

- The portfolio manager will need to carefully screen each mEVM template to spot activity-based milestones. Many project managers will need some kind of mentoring to spot and recast activity-based milestones in terms of deliverable-based milestones with binary completion status.

- After activity-based milestones are replaced, the mEVM template needs to be approved by project owners to ensure that they stand behind the new deliverable-based milestones.
- The problem of choosing activity-based milestones typically goes away once project managers begin to understand mEVM.

Advanced Evasion Tactics

Once the basic concepts of mEVM are understood and the rollout is completed, the true performance of the projects start becoming visible to all. Simultaneously, projects begin to understand how the mEVM system works. Taken together, these conditions give rise to more sophisticated methods of *gaming the system* on the part of the projects, to explain why they are underperforming (some of these are listed in upcoming text). (Notice that the tactic numbering continues from the previous section.)

Tactic #6: Exposing the Use of Padded Estimates

What are padded estimates? It's the practice of building in a *buffer* for your work estimate for your deliverable. Imagine that the following deliverable *build wireframe of new corporate website* is estimated to realistically take about six weeks of work. However, the project manager *pads* the estimate by two weeks and makes the estimate *eight weeks of work.*

What's the benefit of padding? It *hedges* against the variability that goes with every work. If the work actually comes in at six weeks, the project looks like it's over-performing—they got it done in six weeks instead of eight! What if the work actually winds up taking eight weeks? The project still looks like it's coming in on time, just as planned. The practice of padding gives the project a cushion to deal with uncertainty.

What's the disadvantage of padding from a portfolio manager's perspective? It's a huge performance drag on the portfolio. Just consider the project from the previous example, padding the six week estimate by two additional weeks— that's a 33% increase! Imagine if they did this for most tasks—their $1M project would now be seeking $1.3M in funds from the portfolio. Now imagine if all of the projects in the portfolio did this for the deliverables in the portfolio—the portfolio has an efficiency reduction of 33%! Consider all of the projects that could have been started with this money that is *kept as padding.* Also consider that this 33% hides or masks the inefficiency in the projects' execution. Even if the projects execute efficiently, they tend to hold on to the *padding* money, only returning it late in the year when it's too late to start any new projects. How does mEVM work in this situation? With the use of mEVM, padding becomes clear if the project consistently shows a surplus, with the mEVM graph looking as shown in Figure 12.1.

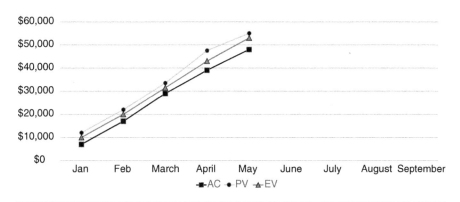

Figure 12.1   mEVM graph of a project with substantially padded estimates

One observation from Figure 12.1 is that the actual cost always runs lower than the planned value and EV. This is a clear indication that the project could safely *turn back* excess funds and continue to deliver as planned. By doing this across the projects, a significant portion of funds is *returned* to the portfolio to be allocated to other purposes.

The aforementioned example shows how mEVM spots padding after a few months of activity. However, there is a way that mEVM can spot padding before the project starts execution. As the organization matures in its experience with mEVM, it is possible to build a catalog of prices for the most common tasks— for example, *building wireframe for a website* or *execute one hundred test cases* should be a standard task that costs about six weeks of labor at the standard rate of the organization's labor cost. Once this is in place, a project that tries to estimate 10 weeks of labor for one of these standard tasks would be spotted and remediated during the mEVM template creation stage itself.

Tactic #7: Blaming Others for Delays

As mEVM works its almost inexorable magic, several projects will start showing up as late or underperforming. When that happens, projects tend to provide a wide range of reasons/excuses. One of these might be: "We're not getting credit for several tasks that are complete except for _____ (insert third-party dependency here)."

Here's a real-life example: A Vice President (VP) of Information Technology was tasked with executing a strategically important project with multiple tracks of activity. One of the first deliverables of the project was to define charters and get them approved by project sponsors, but quite soon the mEVM graph revealed that the project was going nowhere—costs kept going up and the EV curve was barely above zero, signifying that no deliverables were complete (the earning rule was 25/75, which meant they got some minimal credit for starting these charters). By way of explanation, the VP came to the portfolio governance meeting and made the case that the charters were waiting on the desks of executive leadership and that the project had done all they could and hence, deserved to get the EV associated with those deliverables.

What should the portfolio manager do in these situations? The action to invariably take is to hold firm to the mEVM concepts—deliverables only get credit when they are 100% complete, no exceptions. The fact that a deliverable is prevented from completion because it is waiting on someone to do something does not make the deliverable complete—rather, it calls attention to the risk created for the project by that action not becoming complete. The risk should be remediated, not buried by giving EV credit for incomplete tasks.

Tactic #8: We Want to Reset Our mEVM Template and Graph

From past experience, most projects take to mEVM reluctantly. The mindset is along the lines of *do the minimum—like we did for all the other various reporting templates we've been made to fill out in the past.* However, within a couple of cycles, projects realize that mEVM is quite different from the other techniques, in that they are being held to a baseline and this is graphed against real data such as cost and time. Quite soon, it becomes apparent to the project and to portfolio governance if the project is either not doing well or the baseline was not built with enough thought.

This is actually a positive development, in that the project *wakes to reality*—both the reality of their performance and the reality of how that performance is being tracked. This is typically followed by the realization of the need to do the mEVM template again, this time applying a lot more thought and analysis before picking the milestones and their corresponding costs. A portfolio manager, therefore, should not be surprised when project managers and/or their owners want to *reset their mEVM*. The caution here is that some projects may seek to repeatedly reset their mEVM templates in order to *wipe away* the previous mistakes. Here are some pointers as to how to proceed in that situation:

- It is recommended to allow projects to reset their mEVM graph in the beginning, even if only once. This reinforces the flexible nature of the system and also allows for wider adoption. However, it may also be prudent to keep the original mEVM graph for reference, in case there is a need to compare how we are doing from the original projection.
- It must be kept in mind that although the graph may have been reset, the project has the same budget as before. If the new mEVM template shows that the project may need additional money, this is a separate request that the portfolio governance must consider.

- A project that repeatedly fails to perform according to its mEVM template is signaling that there are deeper issues that are troubling the project. Instead of performing periodic resets, the portfolio manager and governance should undertake a deeper analysis and possibly put the project on an EML.

Tactic #9: mEVM Graph Doesn't Tell the True Story

As mentioned before, mEVM is a relentless technique once it's up and running. Unlike the *red, yellow, and green* system of reporting and other subjective measures, it offers no wriggle room or gray area. Projects find themselves in the sometimes uncomfortable situation of having to explain why they are falling behind on cost and time baselines. It's not unusual (at least in the beginning, as the organization understands how to work with mEVM) for some distressed projects to complain that the mEVM technique is too unidimensional or simplistic—in other words, their project is *too complex* for mEVM to capture and represent correctly. This complaint is manifested along these lines: *although the mEVM graph shows we're behind, the graph doesn't tell the true or complete story.*

While there may be some truth to project complexity, it must be kept in mind that mEVM is solely focused on deliverables. It really doesn't matter whether the project is simple or complex—or if the project is strategic or routine. It simply shows how we're doing with respect to our cost and time plan. The arithmetic is inescapable. The challenge of the portfolio manager and portfolio governance is to communicate this to the project owners and project managers. A combination of tact and firmness, coupled with the driver of funding (*must follow mEVM in entirety to get funding*) are usually enough to carry the day.

It may also help to dive deep into the mEVM template—typically the CV and SV are caused by one or two major deliverables that have not been completed, leading to a drop in EV. The reasons why those deliverables could not be completed could become a meaningful discussion that the project may want to engage in (as opposed to arguing about the mEVM graph showing a variance).

Tactic #10: Starting (Noncritical) Activities in Advance to Get Additional EV

In most organizations, project managers are remarkably adaptive in dealing with change. When it comes to mEVM, it takes a couple of cycles before they begin to grasp the levers and moving parts of mEVM. Unsurprisingly, some project managers have been known to try to game the mEVM system in various ways. One of these methods is to *start noncritical activities in advance*. What does that mean exactly? Here is an illustration:

Consider a project with a deliverable (called Deliverable #1) that has $10,000 worth of EV and is due to be complete in the month of January. For various reasons, the project is unable to complete Deliverable #1 in January (although it has incurred the $10,000 cost in working on Deliverable #1). Under this situation, the project would have a negative CV of $10,000—something that the project manager would not like to be shown at the portfolio governance.

What if the project manager could save the day by starting four new deliverables that are based in March/April? According to the 25/75 earning rule, they would get 25% of the EV of these four deliverables and this would (hopefully) cancel out or diminish the $10,000 variance (which was caused by noncompletion of Deliverable #1). This would work great, except that there is no good reason to start new deliverables that are months away while we haven't closed current deliverables. It also creates an untrue picture of the project's true performance and postpones visibility of problems. All in all, this is a short-term manipulation of the mEVM data that serves no purpose beyond obfuscation for the time being.

How should the portfolio manager respond in a situation such as the above? Countermeasures may include:

- Consider instituting a rule that a project cannot claim EV for activities that are more than one month away (Example: If we are in January, a project cannot start activities slated for March or beyond and obtain the EV for those activities).
- During the monthly mEVM review with the project, keep an eye out for activities started that have no seeming dependency to the current month's activities.
- Realize that this is a strategy with diminishing returns—the project will eventually run out of *future activities* to start and will have to come to terms with the current state of the project.

THE ENHANCED MONITORING LIST—
ALSO KNOWN AS THE *KILL LIST*

We've touched upon the EML in numerous chapters in this book. The EML is a sort of *probation* for troubled projects. When a project is revealed, through mEVM data, to be repeatedly falling short of its cost and schedule commitments, it becomes eligible to be designated as an EML project with additional scrutiny and assistance from portfolio governance.

Once projects are able to see how the EML operates, there is a huge incentive to do well and stay off of the EML. This, in turn, creates a resolve to engage deeply with mEVM and work with mEVM data, since the data is the determinant of getting on or off the EML.

Some projects are able to utilize the additional scrutiny of being placed on the EML and turn that into a sense of urgency, which enables the project to come back on track. This works to the portfolio's benefit by enacting speedy rescues of troubled projects which, left unattended, would turn into major sinkholes of the portfolio's resources.

Conversely, some projects have deep-rooted problems that do not lend themselves to resolution, irrespective of the effort. If a project is repeatedly unresponsive to various actions to bring it back on track, it becomes a candidate for termination.

LEVELS OF PORTFOLIO CAPABILITY

Level 1

- mEVM just rolled out or is about to be rolled out in the organization
- Organization still coming to terms with how it works
- Both project managers and portfolio governance members need a lot of handholding
- Some projects have mEVM artifacts in place and some don't
- Some projects still resist adopting mEVM
- Some projects are given a pass from having to adopt mEVM

Level 2

- mEVM rollout complete—all projects following mEVM
- Regular mEVM report outs at portfolio governance meetings
- Some handholding needed for projects with their mEVM setups
- Some projects still trying to reset their EVMs every few months
- Projects are yet to understand how to leverage their mEVM data for their own benefit
- Portfolio governance yet to leverage mEVM data to make key decisions— mEVM not aggregated or used to measure strategic attainment
- EML not yet set up or not yet coupled with mEVM data

Level 3

- mEVM rollout in place for a year or so
- Regular mEVM report outs at portfolio governance meetings

- Project teams need very little handholding from the portfolio team—all project managers understand how to do mEVM
- mEVM resets by projects are rare and driven by external events (as opposed to inadequate first drafts)
- Portfolio governance understands mEVM data and uses it to make key decisions, such as requesting funds giveback from projects
- EML is set up and its functioning is in lock step with the mEVM system
- Projects understand mEVM data and how to use it for self-regulation
- mEVM data is aggregated to produce meaningful information for portfolio governance
- mEVM data is repurposed to measure strategic attainment value

CHAPTER SUMMARY

In this chapter, how an organization should plan to roll out mEVM was discussed in depth. We also looked at short-, medium-, and long-term considerations to keep in mind during the rollout of mEVM—in other words, what a portfolio manager should expect to see in terms of the portfolio's functioning as a result of mEVM. We also covered extensively some of the *evasion tactics* that project managers have been known to use while trying to adapt to the performance visibility created by mEVM. These tactics were accompanied by a list of countermeasures that are recommended for the portfolio manager to take against these tactics. We also introduced the concept of the EML and discussed how it would function with the performance data generated by mEVM. Finally, we rounded out the levels of portfolio capability as measured by mEVM roll-out maturity.

13

WHY mEVM WORKS

INTRODUCTION

We've covered a lot of details regarding modified earned value management (*mEVM*) in the previous chapters in this section. While these are essential to a successful rollout of mEVM in the organization, it is very important to preserve the strategic/big picture view of why we are doing mEVM. It is also good to keep in focus the big reward that comes at the end of a significant effort in rolling out mEVM. This chapter *rounds out* the theory of earned value management (EVM) with the following aspects of the technique:

1. Covers in detail the criticisms of EVM and how our implementation of mEVM works around those valid objections
2. Draws out the distinction between EVM and mEVM
3. Explores why mEVM fundamentally works in its ultimate objective of creating a high-performing portfolio.

WHY EXPLORE CRITICISMS OF EVM?

It's important to explore criticisms of EVM because some stakeholders *may* push back by bringing up these traditional shortcomings of the system and question the need for EVM. So, it's essential to show that these have been considered and mitigated.

THE CRITICISMS OF EVM

While EVM has been accepted as an effective tool for a long time now, several valid criticisms have also been aired for a while. Some of those criticisms include:

- EVM is a huge effort to implement
- EVM is very expensive, needing a complex tool and/or system to support it
- EVM creates a lot of data and needs a lot of resources to crunch the data
- EVM has complex terms and formulae that the rank and file of an organization find hard to grasp
- Organization questions the value of this complicated tool

HOW mEVM OVERCOMES THE TRADITIONAL CRITICISMS OF EVM

While the EVM technique has several valid criticisms, as mentioned before, we need to keep in mind that we're not doing EVM—we are doing a much lighter version, called mEVM, which has specifically been designed to overcome the criticisms mentioned above. mEVM is differentiated from standard EVM in the following ways:

- **Improvement #1—mEVM requires a much lower effort to implement compared to EVM**: By reducing the implementation effort to one based on three simple artifacts, mEVM acquires a huge advantage on the traditional EVM implementation effort.
- **Improvement #2—Low cost to implement**: mEVM uses PowerPoint and Excel—it doesn't need any expensive software or people to administer it. This keeps the cost low and also makes the technique approachable by the rank and file.
- **Improvement #3—Low effort to process data**: Due to the binary nature of deciding whether a deliverable is complete or not, the resource intensive effort of deciding *percent complete* is avoided. With this design feature, mEVM makes it very easy to process the data for each month's snapshot. From experience, mEVM takes no more than 30 minutes per project, per month to update.
- **Improvement #4—Easy to understand format**: mEVM produces the same data produced by EVM such as cost variance (CV) and schedule variance (SV). However, it makes it digestible for decision makers by producing a graph, which is an easy way to understand the health of the project.

- **Improvement #5: User friendly at all levels**: By using simple terms for CV and SV, mEVM keeps the whole system accessible to everyone—it is not complex and/or hard to follow for the project manager or for the executives who consume the data to make decisions.
- **Improvement #6: mEVM can be scaled/aggregated**: mEVM data is stackable, which allows it to form interesting composite views of the portfolio's information. This ability to aggregate allows the portfolio office to switch between the big picture and the details, giving portfolio governance visibility across all dimensions of the portfolio. Aggregation of mEVM also allows visualization, analysis, and trend prediction at various levels of granularity across the portfolio.
- **Improvement #7: mEVM can be used to measure strategy attainment**: Measuring strategy attainment has always been a sought-after goal for portfolio offices, as well as executive leadership. Thanks to mEVM, this is now possible to do in an objective manner. As illustrated in Chapter 11, by making a few design tweaks to the mEVM artifacts and applying them to the task of measuring strategic milestones, it is possible to create a system of measuring strategic attainment in a visually intuitive manner.

THE CANARY IN THE COAL MINE

Beyond all the finer points of which flavor of EVM is best, there is a very clear value delivered by the mEVM technique: it produces a clear answer to the question of whether the project is unfolding as promised, in terms of cost and schedule.

In that aspect, mEVM is the *canary in the coal mine*—it provides early warning that things are going off course in a project long before it becomes obvious to everyone. This provides the most valuable insight to a person managing a portfolio of projects—an ability to intervene and potentially save portfolio funds from being wasted. Some projects are just not feasible to execute, and the sooner we accelerate to this point of discovery the better—because the funds can then be used to support other projects that are shown to be performing well.

WHY DOES mEVM WORK?

mEVM works because it calls out, very clearly, which projects are working and which are not. Consider a portfolio of projects, all reporting using mEVM. Odds are that some of them would report small/no variances in cost and schedule, while others would have significant variances. Quite simply—some projects are doing well and some are not. Let's just call the chronic underperforming projects *bad* projects. Fast-forward to the next month—it could be that some of the *bad* projects are now doing okay, but a few of the *bad* projects are doing even worse than before. Within a few months, it becomes quite clear which projects are *never going to get better*, while other projects are performing average or well. This does two important things:

- Allows the portfolio manager and other decision makers to focus their governance on the *bad* projects, with a view to bringing them back on track
- For *bad* projects that are simply not able to turn around, mEVM provides the portfolio manager and decision makers with the trend information to make a decision on termination

In effect, mEVM works because it provides an exit channel for low-performing projects. Few portfolios have such an exit channel, and this is the primary reason for the mediocre performance of most portfolios.

SECONDARY REASONS WHY mEVM WORKS

As we stated, mEVM brings a massive benefit to the portfolio by removing bad projects. However, experience has shown that mEVM also has a host of secondary beneficial effects. In essence, it elevates the game for the rest of the (non-bad) projects by giving them visibility to true performance as well as early intervention tools. Here are some of the secondary effects observed when mEVM is implemented:

- **Effect #1—Forces the project manager to make a *real* project plan**: It's quite astonishing how many expensive projects, in big reputable organizations, are run without a proper project plan. What does a proper project plan mean? It's not just an Microsoft Project file that lies forgotten in some folder. It means a plan that reflects reality, has key milestones and dependencies among tasks, and is aimed at producing real deliverables at defined and tracked costs. The very first mEVM artifact, namely the mEVM Excel template, cannot be created without a project plan. mEVM provides that wake-up moment for a project—when they realize they have been operating without a plan.
- **Effect #2—Quickly exposes a noncredible plan**: When people are forced to make a project plan, it's almost a certainty that the first plan they produce will neither be realistic nor credible. However, within a couple of reporting cycles, it will become obvious from the mEVM graph that the plan is not based on reality. This provides a second wake-up call to the project to produce a real plan. Knowing that they will be held to this plan

creates the right incentive to think through the deliverables and create a much better second plan. (Some projects will need to create a third plan before it is acceptable.) Why is this so important? Consider—without these *wake-up calls*, the projects would have continued to plod along, with mediocre or disastrous results. A project without a solid plan has a slim chance of succeeding.

- **Effect #3—Changes the politics of reporting:** As covered in Chapter 5, project reporting is very political. In the absence of objective criteria, people manipulate subjective criteria to obscure the true status of the project. The politics tend to be more on the lines of *you have to believe what I'm saying until you can prove otherwise*—and the *otherwise* never arrives until the project is a flaming disaster. When mEVM is adopted, the whole politics around reporting undergoes a marked change. The politics are now structured around *here's why we are not doing well* and/or *this is our plan to reduce the CV and SV by next month*. Given the intense focus on performance, it's only a matter of time until project underperformance becomes politically untenable. In this way, mEVM actually changes the politics to make it work in line with the portfolio goals.

- **Effect #4—Raises the game for everyone:** As people start seeing the effects of mEVM—such as the spotlighting of underperforming projects and possible termination—it creates a huge incentive to manage the projects right and avoid getting on the *bad list*. This, in turn, elevates the expected performance for everyone, since there is now a *race to perform well*. As a result of this, a host of best practices start becoming visible—project planning, forecasting, tracking, and accountability are all tuned up. In effect, mEVM energizes the projects with the tools, visibility, and motivation to prevent their project from becoming a *bad* project.

- **Effect #5—More efficient distribution of resources:** With mEVM, projects that hold on to extra funds can be spotted. These funds can then be allocated to other needs—either other projects that need more resources or to start new projects. Overall, this promotes a more lean and disciplined approach to resource allocation. Remember that prior to mEVM, projects could hoard funds and not even know it. Projects hedge for uncertainty by holding onto funds, and returning it in the very last months of the year when it cannot be put to much use.

- **Effect #6—Promotes proper functioning of annual planning:** As we saw at the end of Chapter 3 on annual planning, the one scenario to avoid would be to bet big on a few projects that wind up not delivering. The better scenario would be to make a balanced bet on several projects, at least some of which may deliver. For that to work, mEVM needs to be in place so that poorly performing projects can be spotted and terminated ahead of time. The goals of annual planning are helped because more projects can be funded and started—with their mEVM performance deciding which ones can continue.

CHAPTER SUMMARY

While the preceding chapters were focused on the tactical implementation of mEVM, this short chapter showed the strategic/big picture view of why mEVM is worth doing. It approached the criticisms of traditional EVM and listed how these criticisms have been addressed by the simplified techniques of mEVM. The chapter addressed the primary and secondary reasons why mEVM works and described how the implementation of mEVM benefits the portfolio and even the larger organization.

Part III

Implementation Strategies for the Real World

14

COMMON PORTFOLIO PROBLEMS
AND SOLUTIONS

INTRODUCTION

There are many well-respected veterans of the field who have written fine texts on portfolio management that are very valuable when starting on the portfolio management journey. From personal experience, one question that the reader is often left with is: "How do I apply these concepts to the immediate problems I am facing as a portfolio manager today?" That question led to the thought that it would be useful to include a list of scenario-based solutions that a portfolio manager could deploy immediately in their organizations. I have also thought it would be useful to have a simple and direct checklist to walk through most portfolio topics to ensure that I have them covered in the real world.

Those thoughts, and the feedback from several accomplished colleagues, form the basis of this section of the book—namely, what happens when portfolio management meets the real world. Throughout this section you will find a listing of the most common obstacles one is likely to encounter as a portfolio manager. You will also find approaches in dealing with these problems, as well as references to earlier chapters in the book that deal with these topics at length.

WHERE DO WE START?

Portfolio management can be overwhelming. The average portfolio has so much opportunity to improve that the most fundamental question facing portfolio managers is: "Where do I start improving my portfolio? Which problem do I solve first?" This is not a trivial question. Every organization is at a different point in its portfolio management journey. And every organization also has their peculiarities that may be unique to that place. Finally, some improvements are not easy to pull off and may not be the best investment of time and effort for the portfolio in its current state. Taken together, they constitute a custom problem to which there may not be a *one-size fits-all* approach. So, the recommended solution is to simply start at the first problem scenario listed in the following section. If that scenario has already been handled in the organization, then simply move on to the next problem. It may still be a good idea to browse all the scenarios because there may be a perspective in the provided solution that the reader could still use to optimize their already-solved situation.

Situation #1: I Have Been Asked to Start a Portfolio Management Function from Scratch. How Do I Go about It?

It can be daunting to start a portfolio from scratch; however, it is also a great opportunity to build it right. Here's how one would go about it:

- First, build a list of projects that are currently being executed. This may require talking to the rank and file of the organization, removing some duplicate names, and some validation/iteration to create a good list.
- This then becomes the *official list of projects* or *the portfolio*. New projects cannot be added to this list without a formal entry procedure (explained in the following text).
- For the projects in this portfolio, collect the most basic and informative attributes such as:
  - Official name of the project
  - Budget of the project (if known) (if not available, see Situation #2)
  - Current actuals, year to date (if known)
  - How much more money each project will take to complete (if known)
- Also collect information from Finance about what the total budget allocated for the portfolio is—this is your portfolio budget.
- As mentioned before, once the project list is solidified, new projects cannot be added without the explicit involvement of the portfolio manager. This would involve filling out a basic intake form (refer to Chapter 2 for some pointers on what to include in the form). The form would then have to be reviewed and approved by portfolio governance, whose function is explained next.
- Share this list and the portfolio budget with *portfolio governance*—the decision makers of the portfolio—and explain that this is the preliminary portfolio. This should be a recurring meeting, preferably monthly.

(What if there are no *decision makers* identified? Great—then you get to suggest who should be informed to make this a broad-based, participatory exercise).

- In addition to the previous information, the following data should be shared each month with portfolio governance:
  - New projects that were started (along with budget information)
  - Projects that were completed (along with final actuals)

This is a very basic portfolio in flight. The journey has only begun.

Situation #2: No One Can Tell Me What the Budget Is for a Project

What happens when there isn't a formal budget for a project? This usually happens when projects are started without much oversight. When too many of these ad hoc projects are started, the expense begins to rack up and now the portfolio manager is tasked with bringing these under control. The challenge still remains to identify a budget for the project when no such formal budget was stated. Here are some pointers on how to proceed:

- If there is no budget for a project, the portfolio manager has to construct a simplified working budget. This could be done in a couple of ways:
  - The project manager needs to provide an estimate of how much it would take to complete the project—this is the standard estimate to complete. This, when added to the money already spent, forms the total budget of the project. (What if there are no actuals? See Situation #3.)
  - The other alternative to creating a budget is to do a bottom-up estimation—namely, modified earned value management (*mEVM*). Although this is a little effort intensive, this forms a more solid number that can be used as a budget. The steps to using mEVM to construct a working budget are outlined in the next few bullets.
- Using the project plan of the project, the portfolio manager needs to build (or guide the project manager to build) the list of major deliverables and their respective completion dates. This is then compiled in the standard format known as the mEVM Excel template, which is covered extensively in Chapter 9.
- The advantage of the mEVM method is that a fairly high quality budget number is obtained, as well as an estimate about what will be spent in each forthcoming month.

Cautions in Situation #2

- Keep in mind that the project owner and the project manager own the budget number for the project and that the portfolio manager is only helping them build it.
- Too often, the project produces an inaccurate budget number—and then the portfolio manager is called out as the producer of the faulty estimate.
- To get around it, the portfolio manager can proceed in the following ways:
  - The easiest way is to ask the project owner/manager to produce the estimate. This way the ownership is clearly established from the start.
  - The next option is to explicitly obtain approval from the project owner/manager about the budget number created with the portfolio manager's help. This documented approval will be useful if the budget estimate is found to be faulty later.
  - In the case of the mEVM option, the preferred approach is to obtain explicit approval of the mEVM Excel template that was constructed with the project team's help. Again, the advantage with the mEVM is that it would only take a couple of months before the budget number is revealed to be faulty or good.

Situation #3: No One Can Tell Me What the Actuals Are for a Project

In some organizations, actuals are not available for projects. Although this seems a little surprising, it may have to do with some decisions taken around implementation of the enterprise resource planning (ERP) system which hides the project costs among other buckets. However, it's quite clear that no portfolio can function well without being able to track actuals by project. Here are some pointers concerning how to start measuring actuals for a project:

- Finance should provide guidance on whether they use project codes (or can otherwise tie spend to a particular body of work). If they can, that's great—the solution may be as simple as a monthly report showing how much money was spent on this project code. However, this works only for purchase orders (POs) and external spend.
- We still need to account for the cost of internal labor spent on a project. If the organization uses timesheets, it should be a straightforward process to run a report that lists the total hours of labor spent against a project.
- Taken together, the POs and internal labor should account for most if not all the actuals of a project.
- What if we don't use timesheets? Things get tougher without the use of timesheets—we would need to get a fairly detailed breakdown of who

worked on the project for how much percentage of their time, and turn those total number of hours into dollars using the standard labor rate. The result would be approximate and not an exact way to get the total cost of the project.

The aforementioned measures are enough to create a temporary solution to obtain (approximate) actuals for a project. In the long term, though, a systemic solution involving changes in the ERP system and the finance/accounting process of the organization is needed. A closer partnership with Finance is also needed in order to ensure that the previously mentioned concerns are taken into account when upgrading the ERP system or implementing fixes.

Situation #4: People Complain that Portfolio Management Is Too Cumbersome

A common situation encountered by most portfolio managers is the complaint from the organization stating that portfolio management is too cumbersome. The feedback is along the lines of: "You're adding too much process and we're not getting things done." The portfolio manager has to tread carefully here. Here are some pointers in navigating this situation:

- Acknowledge the feedback and promise to work with everyone to make the process more efficient. Why bother doing that? Simply because this is going to be a long, drawn-out exercise and the portfolio manager needs the support of the rank and file to complete the journey.
- Recognize where the feedback is coming from. The project owners and project managers are now having to deal with the changes introduced by the portfolio management process and procedures. They are no longer able to start projects at their discretion, run them without oversight, and close them out without an accounting of costs and benefits. The reduction in their *autonomy* creates the momentum to complain against the processes. Sometimes the feedback is just an indirect protest at the loss of power and autonomy that was previously available to the project owners and the project managers.
- Next, consider where changes can be made without diluting the intent or efficacy of the process:
  - Example 1: An analyst from the portfolio team could complete the mEVM forms and get the signoff from the project team instead of insisting that the project team fill out the forms themselves.
  - Example 2: Simplify the project intake form and make it a single sheet only.

- Example 3: Be proactive in informing projects that it's their turn to present at the portfolio governance session that month.
- The bottom line: Establish a sincere partnership with the stakeholders and demonstrate that this is a joint effort—that goes a long way toward achieving participation.
- All of the protests may not be uniform. Some stakeholders may have a genuine issue that can be remediated with some effort on the part of the portfolio office as outlined in the preceding examples. Other stakeholders may be chronic malcontents driven by the loss of their previous freedom. It is important to separate the two categories so that the malcontents are isolated as much as possible.
- Sometimes, people just don't want to play according to the new rules. In those situations, the portfolio manager has to channel the mandate given to them by executive management. Withholding approval for new projects, and/or approval of funding requests, may have to be done to drive home the point.
- A more subtle way of ensuring compliance is to let people fail in full public view. For example, have a stubborn project manager appear at the portfolio governance session and explain to the executives why their project is not meeting the agreed upon portfolio standards.

Situation #5: Bad Projects Are Not Terminated

A common problem across organizations is that bad projects are simply not terminated. For some inexplicable reason, companies are determined to throw good money after bad projects that are highly unlikely to turn around or deliver. In many cases, it might be a political game. Everyone knows the project is a disaster but are simply unwilling to step up and make the much-needed decision to terminate. Historical precedence may play a huge role in the reluctance to terminate a project—*we've never terminated projects* is the mindset that needs to be remediated. Here are some pointers to address the situation:

- This is where strong governance is indispensable. One of the explicit mandates given to the portfolio governance team must be to terminate underperforming projects.
- The portfolio manager needs to provide to the portfolio governance members the data that helps to make up their mind:
    - mEVM data showing the poor return on dollars already spent
    - mEVM data forecasting the likely total cost of the project if it continues to execute in this fashion
- It also helps the governance members with the termination decision when they are provided with the enhanced monitoring list (EML). This

offers an exit ramp and helps them see that some of the projects on the EML are doomed—and that the sooner they are terminated, the better.
- The portfolio manager can also use certain events to force the termination of projects:
  - During annual planning, data can be presented that shows why some existing bad projects should be *below the line.*
  - During portfolio rebalancing, underperformers can be served up as candidates for termination.
  - During approval of new project proposals, a case can be made that the new proposals can only be funded by terminating some underperforming projects and redirecting their spend to the new projects.
  - If portfolio governance is simply *unable to pull the trigger*, that is, terminate projects, the portfolio manager needs to approach the chief information officer (CIO) to make changes in the governance structure or composition.

Situation #6: We Don't Have a Portfolio Management Tool

Some organizations may embark on their portfolio management journey and feel the lack of a tool. While this may seem like a shortcoming, it may actually be a great opportunity. Many organizations have taken the plunge and adopted one of the popular portfolio management tools on the market (at a significant cost) and later regretted it (please refer to Situation #7). Having an inflexible and mismatched portfolio management tool is much worse than not having a tool in the first place. The latter situation can be molded to the portfolio's favor, but the former situation can be severely constricting and prevent the portfolio from developing to its potential. Having said that, here's how to proceed in handling this situation:

- A portfolio management tool doesn't have to be costly, complex, and massive. Everything that a basic portfolio needs can be accomplished with Microsoft SharePoint at a minimal cost.
- As explained in Chapter 15, the key components of a portfolio management tool are the following:
  - An easy-to-use form to submit new project proposals
  - A list with configurable columns to capture key attributes of active projects
  - A simple workflow that can support the movement of items from one stakeholder to the next and keep track of the same
  - A simple repository to look up standard artifacts for each project

- A repository to look up historical governance decisions for each project
- It's a straightforward, low-intensity effort to create the aforementioned functionality, using the out-of-the-box functionality that comes with SharePoint.
- What if SharePoint is not an option? Although SharePoint is now ubiquitous in most organizations that use Microsoft Office, there may be some situations where even that is not available. In that case, the portfolio will have to make do with low-tech options—namely, throwing manpower at the problem.
- The portfolio office may need to be staffed with a sufficient number of entry-level personnel who can carry out tasks such as composing various portfolio artifacts documents and ensuring that these are reviewed and approved by various stakeholders. Other low-tech options are outlined in Chapter 15.
- At some point, as the organization grows, it becomes necessary to invest in a system that can scale the portfolio without disproportionately increasing the size of the portfolio office. This is all that a portfolio tool needs to have.

Situation #7: We Have a Portfolio Management Tool and People Hate It

As described in Situation #6, organizations that have attempted portfolio management may have taken the plunge and paid for one of the well-known portfolio management tools on the market. This typically involves a significant expense as well as a huge effort in implementation. However, even at the end of this expensive exercise, it may also be quite likely that the people in the organization hate the tool and will do anything to avoid using it.

Why does this happen? Probably because the choice of portfolio tool was premature—the organization may not yet have traversed the difficult journey in setting up essential portfolio management capabilities (none of which have anything to do with a tool). When the choice of a tool is made prematurely, the organization gets *locked* into an inflexible software that restricts the rank and file from doing the things that make sense for the organization. What makes the whole situation worse from a portfolio manager's perspective is that the distaste for the tool often extends to the portfolio management process, too—people are resistant to following portfolio processes because these are implemented through the tool. Here are some pointers as to how to deal with this situation:

- Separate out the disliked portfolio software from day-to-day portfolio management. The enterprise portfolio tool can stay as the *official record*, but most of the day-to-day portfolio management tasks can run outside of the tool.

- This defuses the popular ire of the user base and lets them transact their business in a more user-friendly platform such as SharePoint.
- When it becomes necessary to update the *official record*—also known as *the tool that no one likes to use*—the portfolio team can perform the update, which buffers the users from the undesirable experience.
- It does impose an extra load on the portfolio office to make this redundant entry in two systems, but it may be the most expedient thing to do for the present.
- Over the long term, the portfolio team would be well advised to either retire the enterprise portfolio software or modify/upgrade it to a version that is more acceptable to the users.

Situation #8: Portfolio Manager Has No Power— "No One Listens To Me"

One of the maddening experiences for a portfolio manager is to realize that they may be quite low on the totem pole of the organization. In other words, they just don't wield enough power for people to listen to them and change the way projects are run. The irony is that the portfolio manager can enable projects to run better and deliver significant value to the organization, but sometimes no one wants to give their ideas a chance to work. Therefore, few things change and the status quo dominates. Here are some pointers on how to handle this most fundamental roadblock:

- Initially, the portfolio manager needs to be able to describe and sell a vision to upper management that shows the strategic value of a well-run portfolio. A successful portfolio manager needs to possess a fair amount of salesmanship in addition to technical skills.
- After securing executive buy-in, the portfolio manager needs to leverage that mandate to drive the first wave of desired change in behavior and processes.
- The portfolio manager then needs to demonstrate the benefits of the first wave of changes—perhaps a better intake, better visibility, and better management of project benefits? The aim is to establish a track record of continuous improvement for the portfolio journey. This reinforces the portfolio manager's credibility when they approach executive management for the mandate to drive additional change.
- If done successfully, the portfolio manager eventually becomes a credible, respected voice and is sought for input by the decision makers. The portfolio manager needs to then use that credibility to push for small

incremental changes that cumulatively have a large effect in taking the portfolio where it needs to go.

Situation #9: We Don't Do Strategic Planning

Throughout this book, we've emphasized the portfolio as a vehicle to drive strategic change in accordance with the strategic road map. But here's a basic problem—what if the organization has no strategic plan? Or, as is much more common, what if there is no realistic or actionable strategic plan? Most organizations do have some kind of strategic plan, but it often has no bearing to action on the ground. In other words, there is no connective tissue between the lofty framework of strategic goals and the tactical actions of projects and programs. How can the portfolio manager bridge this divide?

- The portfolio manager can create a faux strategic plan as follows:
    - Define a few strategic imperatives—these can be generic, such as: growing revenue, increasing market share, increasing profitability/efficiency, or building strategic capabilities.
    - Decompose these strategic imperatives into smaller, more specific strategic priorities that are relevant to the organization.
    - Align the current projects and programs to the strategic priorities—use the strategic decomposition table outlined in Chapter 2 or some other similar format to represent the alignment graphically.
    - Present the previously mentioned version of the strategic landscape to the decision makers, including portfolio governance. This may spark some welcome dialogue that results in meaningful changes to the strategic road map.
    - The ultimate end product here is a strategic plan that is recognized as the official plan to be used as a compass for deciding what projects and programs to pursue.
- Sometimes, the aforementioned effort may meet with resistance, with some stakeholders questioning the mandate of the portfolio manager to create a strategic plan. In those situations, the portfolio manager should defer to the people who are wanting to create a strategic plan and focus instead on the end goal—to have a strategic plan and use it to decide what projects to execute. Another option would be to secure a mandate to define a draft plan with the understanding that key decision makers would then use the draft plan as a starting point to make changes and arrive at the final strategic plan.

Situation #10: Our Portfolio Process Has No Connection with Strategic Planning

As discussed in the previous situation, one of the common difficulties faced by a portfolio manager is to navigate a portfolio in the absence of a strategic plan. This situation covers a closely related problem—often, there is indeed a strategic plan in place, but one which has no connection with the portfolio. It may seem like the two activities—portfolio management and strategic planning—are run on two independent tracks with no possibility of intersection. How does the portfolio manager remedy the situation?

- This situation, while quite serious, is better than the previous one where there was no strategic plan to start with. Having a plan eliminates the need for the portfolio manager to create one. It also avoids all of the political wrangling about the mandate of a portfolio manager to play a role in strategy formulation.
- Starting with the official plan, the portfolio manager proceeds to decompose the high-level strategic imperatives into more tightly defined strategic priorities that are relevant to the organization.
- Then comes the exercise to map the current portfolio of projects and programs to the strategic priorities.
- The portfolio manager should expect to see some projects that do not align well to the strategic plan. In that case, a recommendation should be made about what to do with those projects. In most cases, we would want to complete those projects to avoid a complete write-off of the already spent funds for those projects. In a few situations, it may be worth considering a termination of the projects if the money spent is not significant and/or the project's objectives are nowhere near the strategic priorities.
- With the previous exercise, it should be expected that some dialogue may occur about the need to change the strategic plan. This is because the visibility created by the exercise now enables people to clearly see these two complementary entities—the portfolio and the strategic plan.
- Having done the previous exercise for the first time, the portfolio manager needs to ensure that the alignment continues going forward. Some of the ways to do that are:
  - Push for portfolio representation (ideally co-ownership) of the strategic planning activity.
  - Ensure that the portfolio annual planning process relies heavily on the strategic plan (see Chapter 3). Push for continuity of a strategic plan because a plan that changes every year is almost useless.
  - Ensure that portfolio intake considers strategic alignment (see Chapter 1).

- Ensure that portfolio tools such as mEVM also report on the strategic attainment and progress made according to the strategic road map.
- Provide these reports to portfolio governance and make recommendations.

Situation #11: People Want to Tie Their Projects to High-Priority Strategic Programs to Achieve Funding

Once the basic discipline of portfolio intake management takes hold, other kinds of problems begin to emerge. Now that people are prevented from starting projects willy-nilly, they try to align themselves with established, strategic programs in order to get past the hurdle of intake management. How can this be addressed?

- The program manager of the high-priority program needs to agree that this particular project belongs to their program and is included in the program budget.
- The portfolio office needs to enable this signoff from the program manager in one of the following ways:
  - As a low-tech solution, the program manager sends an e-mail approving the inclusion of the new request in their approved program. This e-mail can be added to the supporting documentation for the new request.
  - As a more high-tech solution, a workflow in the SharePoint intake list routes requests to the correct program manager when the corresponding program is chosen in the drop-down. The program manager needs to approve in SharePoint and then the workflow reverts to the portfolio office.
  - The systemic, high-tech approach is always preferable because it takes all unnecessary interaction out of the loop, with less chance for errors, too.

Note that there is a possibility that the program manager may not keep track of how the incoming projects add up to a sum that may exceed the program budget. It is prudent for the portfolio office to keep a running total and raise the flag when the incoming requests cumulatively begin to approach the program budget.

Situation #12: People Start New Projects without Informing the Portfolio Manager

In organizations that are yet to implement a portfolio, it's common to see projects being started without prior communication or approval. Due to organizational inertia, this practice may continue even after the portfolio rollout. What options does a portfolio manager have to curtail such behavior and regain control over project initiation?

- The portfolio manager needs to partner with Finance (see Chapter 19 on how to partner with Finance) and request their help in making sure only POs against authorized projects are approved. This would mean POs that are charged against unauthorized projects will *not* be approved. It's better to make this a systemic implementation such that the financial ERP system would automatically reject such POs.
- The portfolio manager should not make exceptions to these projects that were started on the sly. For example, these projects cannot be presented at the governance meetings until they go through intake.
- Additionally, the portfolio manager should approach the project owners and ascertain why the project was started without notifying the portfolio team.
- Finally, the portfolio manager should recommend that there should be some kind of official disapproval made known from the CIO's office toward stakeholders who persist in starting projects without due authorization.

Situation #13: People Want to Misrepresent the True Status of Projects

One of the most persistent problems faced by portfolio managers has to do with project owners trying to misrepresent the true status of projects. This is typically done to buy time for troubled projects in the hopes that the project manager can turn things around. The project owners may also believe that the problems in the project may work themselves out. Sometimes the project owners are quite unaware of the true status and pass along their incomplete or false understanding of the status to the portfolio. What can be done in such a situation? Here are some pointers:

- As covered extensively in Chapter 5, the portfolio manager needs to understand that the traditional methods of reporting status—for example: red, yellow, and green (R/Y/G)—are simply too subjective to provide a precise readout of project status.
- The portfolio manager has to push the projects to start using mEVM. The rollout has to be handled with care as detailed in the sections dealing with mEVM rollout. Plenty of handholding may be necessary and the portfolio team may have to take the lead in building some of the artifacts for the project.

- Once the artifacts are built, mEVM is relentless in providing a true picture of performance regarding both cost and schedule. If the project is underperforming, this is brought out in full view, often causing project owners to realize the extent of the problem in their projects.
- Once alerted, the project team may want to rebaseline their project and try to get the project back on track. Once again, mEVM will keep the spotlight on the project and inform governance if the turnaround efforts are working and forecast what the cost of the project will be at the end.
- It may even turn out that the project is in serious trouble and needs to be placed on the EML.

Situation #14: People Want to Evade Measurement of Project Benefits

Project owners promise the moon when seeking funding, but sometimes become hard to find when it's time to measure the actual benefits delivered by the projects. One common tactic is to claim that the benefits will only be delivered at the end of the project, at which point the project's money is already spent and there is little recourse. How can the portfolio manager make projects accountable for their promised benefits?

- The portfolio manager needs to start nudging the organization toward following a benefits realization process. The building blocks, as detailed in Chapter 7, are outlined here:
  - The first step in measuring benefits is to create a standard taxonomy in describing benefits. All projects need to use the standard taxonomy in describing their benefits. Projects that refuse to do this need to be sent back at the governance review with a direction to come back with the benefits in the standard format.
  - The next step is to make it mandatory for projects to declare the earliest time that benefits can be produced by the project. Projects that declare that benefits can only be delivered at the end of the project should be treated as high risk investments where nothing can be done if the project does not produce benefits as promised. Instead, projects that can start delivering benefits earlier should be emphasized for funding consideration.
  - The next step is to constitute a benefits review council that meets regularly and reviews the benefits of all projects. A key component of this review is to compare the current benefits with the promised benefits and take action if there is significant variance between the two.

Situation #15: People Don't Want to Follow Process

Once the building blocks of the portfolio management process are implemented, the next major problem for a portfolio manager is to contend with people who want to be granted exceptions to the process. For example, some project owners may want to bring a PowerPoint in lieu of filling out the intake form. Others may want to use their own status reporting system—such as the notoriously ineffective R/Y/G system. How should the portfolio manager deal with this situation?

- In the beginning of the portfolio journey, it may be a good idea for the portfolio office to relax the rules and allow people some latitude in following process and using the official templates.
- However, as time passes, the portfolio office needs to slowly increase the rigor associated with the process. When it comes to people who will not follow the process, a couple of options are available—one is to *not transact*, that is, the portfolio will simply not entertain requests that do not follow process. In other words, a project owner who will not fill out the project intake form will not have their project considered at the intake meeting. The other option is to let the project *fail in full public view*—for example, letting the project owner explain to the governance committee why they are not able to produce performance data for their project in the format that the governance committee has come to expect from all projects.
- In summary, the portfolio office is able to drive compliance by controlling the portfolio resources, primarily funding. If a project's stakeholders refuse to follow process, they may find themselves locked out of funding and other portfolio resources. The portfolio office needs to have (and confirm) backing from executive management before refusing to engage with truant stakeholders.

Situation #16: Projects Are Resistant to Adopting mEVM

As covered extensively in the preceding chapters, mEVM can be a very powerful tool for the portfolio decision makers. In order to get to that point where mEVM can provide insights, projects must first decide to implement mEVM and start reporting their project status in the mEVM format. The stumbling block here may prove to be the projects that refuse to adopt mEVM, thus preventing the portfolio office from taking a closer look at how the project is truly performing. How should the portfolio manager deal with this situation?

- The portfolio office needs to roll out mEVM in a gradual manner, as recommended in Chapter 12. During the initial days of this rollout, the portfolio office needs to go the extra length in helping the organization adopt the new system of measurement. This may involve extra effort on the part of the portfolio team in terms of handholding the project team as they complete mEVM templates and other artifacts.

- As a result of roll-out efforts and the helpful partnership with the portfolio office, it can be expected that quite a few projects would willingly take to mEVM. This isolates the resisting projects that feel growing pressure when asked to present their status at portfolio governance sessions.
- As a last measure, once most projects have adopted mEVM, the holdouts will have to be informed that mEVM is the reporting standard for the portfolio and that projects without mEVM implementation cannot be funded for the next year.

Situation #17: People Complain that the Portfolio Website Is Too Hard to Navigate

As the portfolio grows and the portfolio office deploys more process and artifacts, one predictable effect is the growth of records and documents and the need to keep everything organized. This usually takes the form of a portfolio website, where all the materials and historical records related to the portfolio are stored for reference by the organization. The unintended side effect of a large and growing website is that people find it hard to locate what they're looking for. This frustration may have serious effects if people are unable to follow portfolio process or if their dissatisfaction with the website turns into dissatisfaction with the functioning of the portfolio office. How should the portfolio office handle this situation?

- The primary factor to consider when designing a portfolio website is that most stakeholders have only a basic need or transaction that they need to carry out on the website. In other words, a successful portfolio website is one that enables the vast majority of its users to perform their actions quickly and efficiently. Although this sounds simple, this can be problematic because everyone may have a slightly different simple transaction to execute and it could be hard to optimize the website to accommodate everyone's perspective.
- One tool that can help overcome this situation is the mind map. The fundamental premise of the mind map is that everyone starts from a common place and are able to quickly navigate to the specific task that they need transacted. Figure 14.1 shows the mind map of a typical portfolio office. Using this mind map, any stakeholder in the organization can quickly navigate from the central topic to the relevant main topic

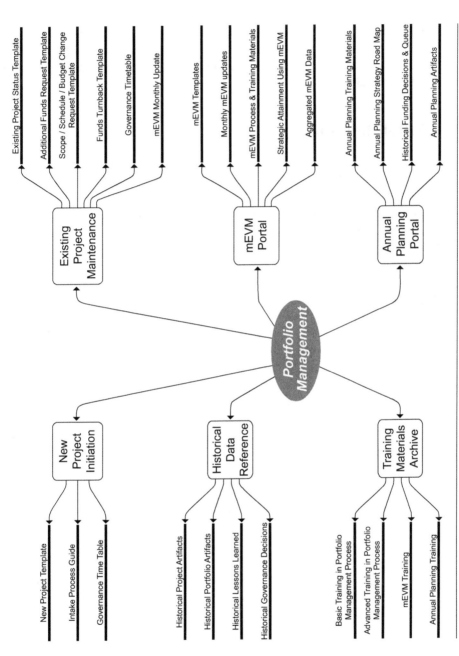

Figure 14.1 Mind map navigation guide for a typical portfolio website

and then finally to the subtopic of interest, which typically has a hyperlink that takes the user to where they need to go. (Chapter 22 has an extensive treatment of the mind map implementation for a typical portfolio website.)

CHAPTER SUMMARY

In this chapter, we explored all of the common problems faced by portfolio managers as they attempt to implement portfolio management best practices. The solutions explored in response to each of these problems consist of a focused approach in drawing content from the other chapters in this book. Portfolio managers are encouraged to go through all of the problems and solutions listed here, as there may be insight that could prove useful to the portfolio manager at some point in their implementation journey.

15

PORTFOLIO INFORMATION SYSTEMS

INTRODUCTION

Portfolio information management systems are among the least favorite tools in any organization. This is a huge problem because people tend to avoid using what they dislike. In turn, this has a debilitating impact on compliance with portfolio management policies and procedures. This chapter explores how to set up portfolio systems and tools that actually deliver results by covering the following aspects:

1. Exploration of why most portfolio management tools fail to meet expectations
2. Discussion of the characteristics of a successful portfolio management system
3. Discussion of the intake process and its system needs
4. Discussion of the annual planning process and its system needs
5. Discussion of the portfolio transactions archival/retrieval system
6. General best practices for portfolio tools

THE REASON WHY MOST PORTFOLIO MANAGEMENT TOOLS FAIL

Before delving into the set up and operation of portfolio tools, it is necessary to have a solid understanding of why most portfolio management tools fail. This is perhaps best explained through an analogy. Imagine a person taking up a new sport—say, tennis. In their newfound enthusiasm, the person buys all of the expensive accessories related to the sport, including a very expensive racquet. However, the individual doesn't have the time or the discipline to take tennis lessons. Subsequently, the odds are rather high that he or she would be disappointed with his or her performance. However, instead of a realization about the need for tennis lessons, most people are likely to blame their racquet and make plans to invest in an even more expensive racquet.

That scenario also plays out in a very similar way for portfolio management. An organization is typically advised of the need for portfolio management after disappointing project execution results. But, rather than build the essential capabilities as described in this book, organizations often find it more appealing to *go buy a solution*. That almost never works because of the following reasons:

- No tool can fully control the actions and decisions taken in the real world by people.
- The workflow of the tool may be a poor fit for the organization's process flow.
- The tools that vendors' business models typically depend on require quick implementation—which is not conducive to the extensive training and change management needed for a successful launch.
- An organization has a finite change management capacity—this may not be able to accommodate *both* a new tool and new portfolio management processes in the same time frame.
- An organization that is evolving its portfolio management process will find its evolution stifled by a tool adopted early in its journey—because the tool only permits what is contained in its feature set.

WHAT A SUCCESSFUL PORTFOLIO SYSTEM LOOKS LIKE

A successful portfolio system is actually a combination of two things that work in tandem:

1. A collection of robust, effective portfolio processes
2. An easy-to-use information tool that integrates with the aforementioned collection of processes and enables the execution of these processes

All of these processes are explained in the sections below, followed by an explanation of how the tool makes the process work in an efficient way. As far as the tool is concerned, SharePoint is the tool of choice that is explained in detail in this book. Why SharePoint? Here are the reasons:

- **It's ubiquitous**: almost everyone is familiar with SharePoint in their workplace.

- **It's affordable**: most organizations have already paid for SharePoint as part of their Microsoft licensing arrangement. Even if that's not the case, SharePoint is very attractive from a cost point of view, especially when compared to the significant investments needed for specialized portfolio software.
- **It's easy to use**: unlike other software, the learning curve is very brief, if not zero (user base may already be using SharePoint software for their day-to-day tasks).
- **It's scalable**: most of the portfolio management functionality can be met with out-of-the-box functionality. In addition, SharePoint can be extensively customized to meet specialized workflow requirements.
- **It's integrated**: SharePoint is extensively integrated with other Microsoft offerings such as Word, Excel, and PowerPoint. In addition, SharePoint is well integrated with Microsoft Project, the most popular project software used by project managers.

PORTFOLIO PROCESSES

As mentioned in the previous section, success in portfolio management systems consists of enabling success in the major portfolio processes. This section explores each of the critical processes that need to integrate with the portfolio management system or tool.

PROCESS #1: INTAKE PROCESS AND TOOL

Managing the intake of new work is central to the functioning of any portfolio. As was covered in Chapter 2, the intake process regulates how new work entering the portfolio is collected, assessed, socialized, and ultimately funded, if found to be in alignment with the mission parameters of the portfolio. A successful intake process needs to have the following characteristics:

- It should be easy for projects to submit a proposal to start a project.
- It should be as brief as possible, while collecting all the information necessary for portfolio governance to make the right decision about approving or deferring the proposal.
- It should allow for different pathways based on the type of proposal. For example, it should be possible for a previously known proposal (such as a new project in an already approved program) to be processed very quickly. At the same time, it should also ensure that a completely new proposal is vetted thoroughly before committing portfolio funds.

- The process should be transparent, showing the status of proposals to all interested parties.
- The process should be integrated with other portfolio management processes.

How SharePoint Addresses the Needs for Portfolio Intake—Basic Mode

Although several solutions exist in SharePoint to provide the aforementioned functionality, the simplest way is to create a custom list to store incoming proposals, as shown in Figure 15.1. Here are the columns that make up the intake list:

- **Column #1—Proposal ID:** The proposal ID is an alphanumeric identifier that can be used to identify the proposal without ambiguity. When the number of proposals becomes large, it becomes impractical to refer to a proposal with its name. This is where it's useful to have an ID to refer to the proposal.
- **Column #2—Proposal Name:** This captures the name of the proposal, and if approved, this would become the name of the project.
- **Column #3—Parent Program Name:** Proposals are often aligned to existing programs. It helps to capture and display that information because these program-aligned proposals can then be expedited through intake. This field has a drop-down with values showing valid program names from the annual plan for the year. *No Program* is also a valid choice, because some proposals are freestanding and do not align to a program.
- **Column #4—Approved in Annual Planning?:** As mentioned in Chapter 3, many projects are seen and approved as part of the annual planning exercise, but may fall *below the line* due to lack of funds. During the intake process, it helps to display the fact that a proposal was already seen and approved as part of annual planning—this could expedite the processing of that proposal.
- **Column #5—Brief Description:** This column contains a brief description of the proposal, which enables the governance body to identify the context of the proposal.
- **Column #6—Aligned Strategic Priority:** This column specifies which strategic priority the proposal is aligned to. This field has a drop-down with values showing valid strategic priorities from the official strategic plan. *No strategic priority* is a valid choice too, because some proposals may be tactical in nature and not aligned to any strategic priority.

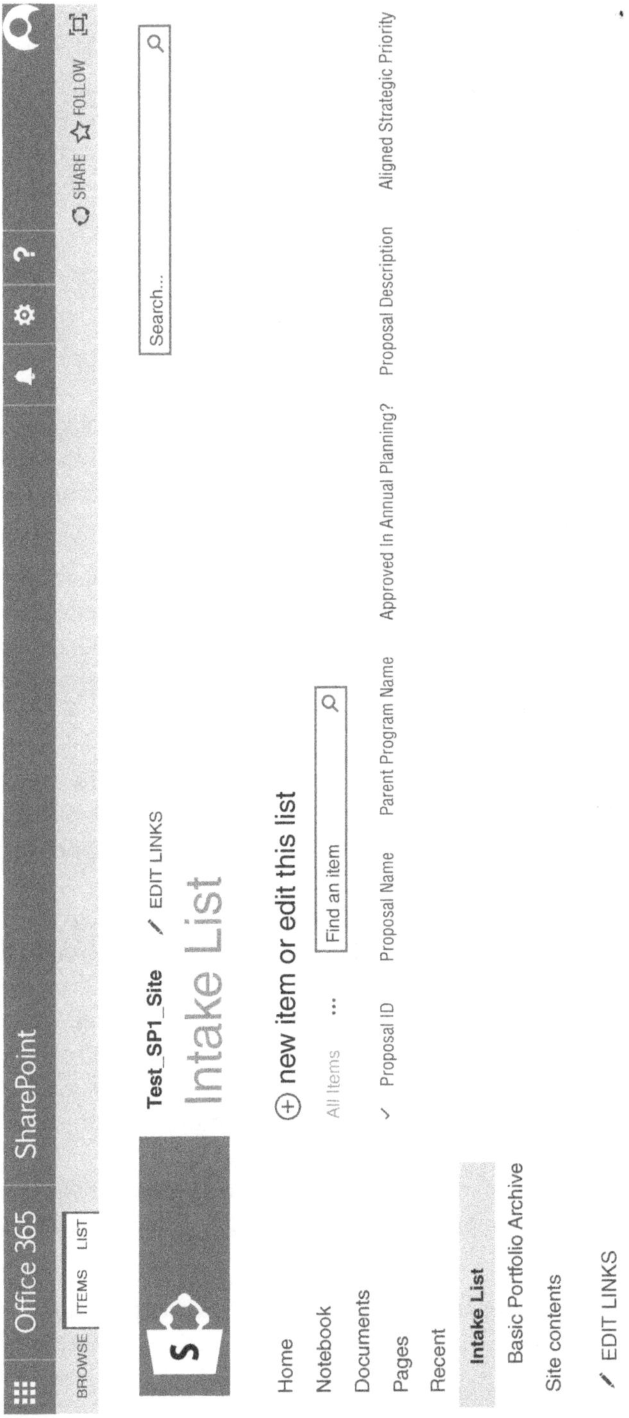

Figure 15.1 SharePoint custom list to manage portfolio intake

In order to enter a proposal into this list, stakeholders are given a *new item* hyperlink that when clicked, points to the form shown in Figure 15.2. Once the form is filled out (see sample entries previously mentioned) and hit *submit*, the proposal turns up as a new entry in the list, as shown in Figure 15.3. It's also worth noting in Figure 15.2 that there is an option to attach a file to the proposal, which may be useful to include attachments that describe the viability or importance of the proposal beyond the brief description field. This is a very simple setup and can be done with any version of SharePoint.

How SharePoint Addresses the Needs for Portfolio Intake—Advanced Mode

The previous section described a basic setup for portfolio intake using Share-Point. For organizations looking to add more functionality to the intake process, the following features can be implemented to make the setup even more effective and streamlined:

- A workflow could be implemented that routes the proposal to various roles within the organization in order to validate the content of the proposal. For example, the proposal shown in Figure 15.2 claims to be aligned to *Strategic Priority #2*. However, only the priority advocate for *Strategic Priority #2* can validate that claim. The workflow could be set up such that proposals claiming such alignment get routed to the corresponding priority advocate for their review. The priority advocate would then record their consent/approval to link the proposal to Strategic Priority #2, which would enable portfolio governance to expedite processing of the proposal.
- Similarly, only the program manager of the *2020 Market Win Program* can validate the alignment of the aforementioned proposal to that program. Once again, the workflow would make this possible by routing the proposal for approval by the program manager connected with the chosen program. The program manager would then record their consent/approval (signifying that the project belongs to the chosen program), which would enable portfolio governance to expedite processing of the proposal.
- If these validations are complete, it should be possible for the portfolio team to recommend these for approval by the portfolio governance team.

Note that a Sharepoint programmer may be needed to set up work flows to work in the manner mentioned here, depending on the version of SharePoint in use.

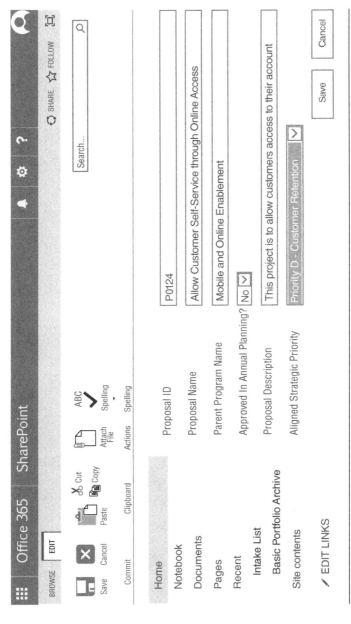

Figure 15.2   SharePoint form to enter new item for portfolio intake

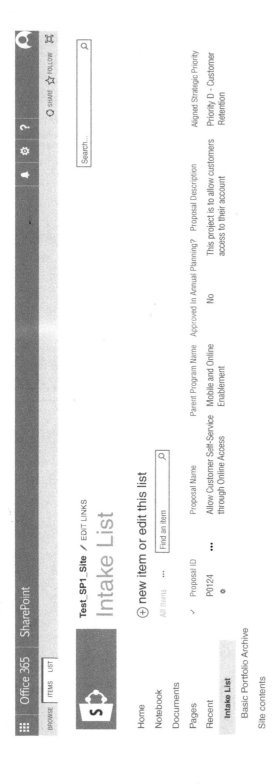

Figure 15.3   SharePoint portfolio intake list with new proposal entered

PROCESS #2: ANNUAL PLANNING PROCESS

We covered annual planning at length in Chapter 3. As previously explained, there are several *passes* that need to be performed to gather all the templates involved in annual planning. SharePoint makes this activity manageable and user friendly, while still gathering all the data needed to make a meaningful decision. SharePoint can facilitate annual planning in either the basic mode or the advanced mode. Both of these modes are explained in the following sections.

How SharePoint Addresses the Needs for Annual Planning—Basic Mode

At a basic level, SharePoint can enable annual planning in the following way:

- The roll out of annual planning (also known as Pass One) can be Share-Point-based—training materials and templates can simply be posted on SharePoint for easy reference by the organization's rank and file. This is important because in most organizations, people are simply unable to grasp all the nuances of annual planning in one or two training sessions. They always find it necessary to *go look up* the training materials as well as the templates when the deadline for annual planning submissions approaches. Having the materials posted on SharePoint provides confidence to the team that they can retrieve the materials when needed.
- SharePoint can be the *holding place* for all of the annual planning templates. Annual planning can involve a large number of proposals being sent back and forth as the various stakeholders in the organization try to arrive at the right message and financials to convey to the decision makers. Sometimes, the proposals are revised multiple times before submission. Trying to get all this done through e-mail is a recipe for confusion and chaos. A cleaner, more elegant option is to ask the stakeholders to upload their proposal to a page on SharePoint. Apart from removing the bottleneck of the portfolio team working through hundreds of submissions, this also enables version control of proposals that need to be revised. From a process efficiency perspective, allowing all proposals to be uploaded to a common repository facilitates lateral dialogue among project owners that is such a necessary feature of annual planning. Through project discovery, stakeholders can discover other proposals to which they may have a dependency.

How SharePoint Addresses the Needs for Annual Planning—Advanced Mode

As explained in the basic mode of facilitation for annual planning, SharePoint can be very effective, even when just functioning as a *holding place* for annual planning artifacts. Also, this can be done with almost any version of SharePoint—even from a decade ago. However, with the advanced features built into the more recent version of SharePoint, it's possible to get dramatically better support from the platform in aid of the annual planning activity, as explored next.

Consider the prevalent mode in which annual planning occurs—a large number of templates (typically Excel) are generated and then these are turned into raw data that is analyzed using bubble charts or some other similar attribute-weighting mechanism. The process of turning the Excel template's fields into raw data can be painstaking—sometimes mitigated by Excel macros and/or Visual Basic for Applications. The whole process is clunky and held together with a lot of manual effort.

Now imagine that many of these steps were rendered unnecessary by eliminating the Excel template altogether. In its place, we have a small number of well-crafted, logical web forms that gather the essential data once and then is directly used for analysis to prioritize projects. Let's be clear—this isn't new by any means. Such a web-based front end supported by a database at the back end was always possible, now and before; however, it needs significant effort and a level of investment typically associated with a major web project.

The major advantage with SharePoint is that all of these formerly expensive components—a flexible, intelligent, modular front end coupled with a robust back end, such as SQL Server—come out of the box in a user-friendly package that can be assembled in weeks, not months. Once the other features are taken into account—such as routing, integration with Microsoft technologies, and other value additive features present in the more recent versions of SharePoint—this solution is hard to ignore in terms of being the logical platform for portfolio management.

Note that a Sharepoint programmer may be needed to set up the fields and reports to work in the appropriate manner, depending on the version of Share-Point in use.

PROCESS #3: ARCHIVAL AND RETRIEVAL OF PORTFOLIO DECISIONS

A typical portfolio generates a substantial amount of data in the course of its normal operation. Consider a portfolio with, say, 10 projects. In the course of a year, assuming that each project comes in for a portfolio governance review just once a quarter, it generates about 40 portfolio decisions (10 projects times four visits). In about three years, this modest portfolio setup yields 120 decisions. Most portfolios have more than 10 projects, which means that the number can only go up. All of these portfolio decisions and the surrounding details around each of these decisions need to be archived and possibly retrieved for later review.

In addition to the previously mentioned data, the monthly modified earned value management (*mEVM*) artifacts need to be stored each month. As explained in previous chapters, mEVM adoption is significantly improved when stakeholder effort is minimal. To keep the effort minimal, stakeholders have to be provided the option to retrieve the previous months' artifacts, to which they can make small, incremental updates for the current month and create the current month's mEVM update. SharePoint can meet this need in both the basic and advanced mode.

How SharePoint Addresses the Needs for Archival and Retrieval—Basic Mode

In the basic mode, SharePoint meets the needs for archival and retrieval through a set of folders as shown in Figure 15.4. Each folder contains the portfolio decisions and transactions that occurred in that calendar year. To allow further navigation, each folder opens up into a set of subfolders representing the months of the year. For example, when the 2016 folder is double-clicked, we see the subfolders as shown in Figure 15.5.

Within each of these folders, there would be a decision letter listing all of the decisions taken in that month. This setup can also meet the storage and retrieval needs of mEVM artifacts. Within each monthly folder, there would be another subfolder for each project, containing all of the mEVM artifacts for that project, updated for that month (this arrangement is not shown).

Although the folder solution is simple and takes minimal effort to create and maintain, it has a couple of major disadvantages. The user has to know with certainty which month contains the information that is being sought. If this is not known (and it's often unknown), the only option left is to open several folders one by one until the desired information is found. This can be quite frustrating for users.

The other problem with the folders is that this is not a scalable approach *over the years* for a portfolio. Considering that each year of portfolio operation creates twelve monthly subfolders, there could be too many folders to manage within just a few years. Looking through all those folders can be a daunting task. The solution is described in the next section with a more advanced SharePoint-based solution.

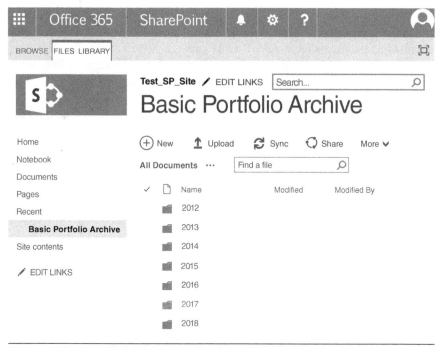

Figure 15.4   SharePoint folder solution for archival and retrieval

How SharePoint Addresses the Needs for Portfolio Decision Archival and Retrieval—Advanced Mode

To overcome the limitations of the simple solution using folders, we define a custom list, called the historical portfolio repository (HPR) as shown in Figure 15.6. Here are the columns that make up the intake list:

- **Column #1—Project Name**: This captures the name of the project that came to the portfolio governance review session and about which a decision was taken.
- **Column #2—Decision Date**: The actual date of the portfolio governance review when the decision was taken.
- **Column #3—Type of Visit/Decision**: A project can come to portfolio governance review for many different transactions. It could be a scope change, schedule change, funds allocation, funds release, general review or funds turnback. This column specifies the exact purpose of the visit. It needs to be noted that there can be more than one type of transaction in a single activity.

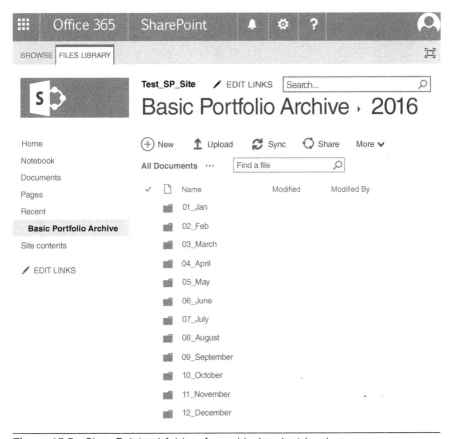

Figure 15.5   SharePoint subfolders for archival and retrieval

- **Column #4—Decision Month**: This column contains the calendar month when the portfolio review occurred and the decision was made.
- **Column #5—Decision Year**: This column contains the calendar year when the portfolio review occurred and the decision was made.
- **Column #6—Decision Outcome**: This column captures the actual decision taken, which could vary between approval, contingent approval, deferral, and rejection.
- **Column #7—Decision Details**: This column captures the details of the actual decision taken, which could contain nuances that need to be elaborated. For example, a decision of *contingent approval* would need some details specifying what the decision is contingent upon.
- **Column #8—Notes**: This column allows the inclusion of notes that may be useful to refer to at a later point.

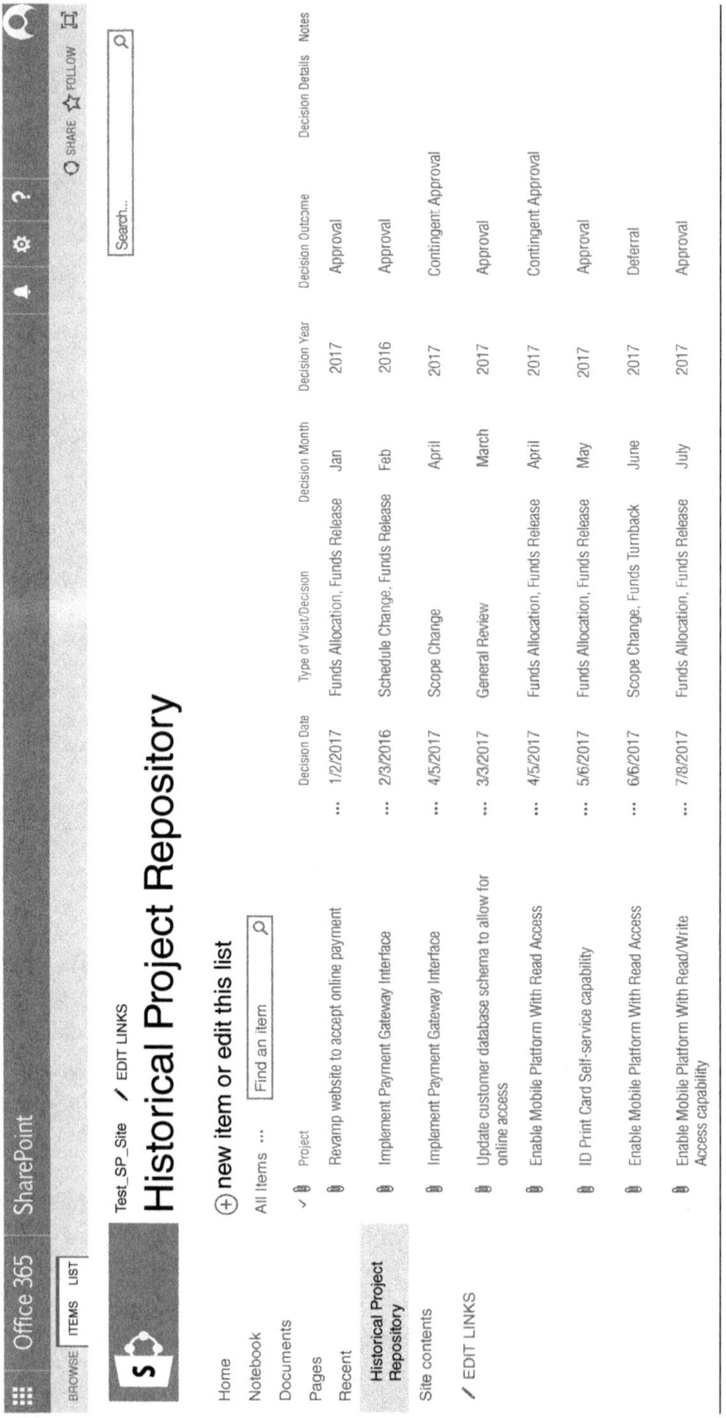

Figure 15.6 SharePoint custom list to manage the HPR

In order to enter a transaction into this list, the portfolio office staff uses the *new item* feature of the custom list, which when clicked on, points to the form shown in Figure 15.7. Once the form is filled out (see sample entries) and the submit button is clicked, the transaction turns up as a new entry in the list shown in Figure 15.6.

How does this custom list overcome the limitations of the folders? Here are some scenarios:

- Suppose there is a need to look up all of the previous historical decisions regarding a particular project—*Implement Payment Gateway Interface*. With the custom list, the user only has to click or hover the mouse over

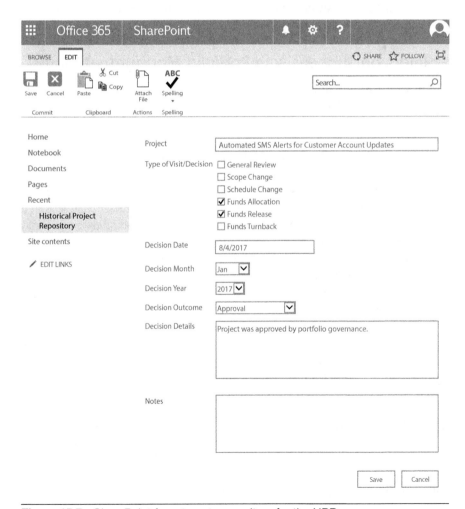

Figure 15.7   SharePoint form to enter new item for the HPR

the Column 1 heading *Project Name* to bring up a filter where this project can be selected.

- The list then shows only the transactions related to the project *Implement Payment Gateway Interface*.
- Suppose that this project has dozens of decisions over the past several years; however, we are only interested in the funding allocation decisions. In that case, the user has to click or hover the mouse over the Column 3 heading *Type of Visit/Decision* to bring up a filter where *Funding Allocation* can be selected.
- The list then shows only the transactions related to *Funding Allocation* for the project *Implement Payment Gateway Interface*.
- In the previously described manner, information can be sliced and diced until the exact type of decision is isolated and retrieved.

Pointers on Success with Portfolio Management Tools

- The portfolio management tool itself is not a problem; it's *when* the tool is introduced and *on top of what maturity/existing portfolio management capabilities* that makes all the difference.
- Some organizations have big enterprise tools already in place, but the portfolios may have a lot of room for improvement. Rather than make changes to the enterprise tool each time there is a change to the portfolio process, it may be best to leave the enterprise tool as it is and transact all the work using a simpler tool such as SharePoint. If the enterprise tool is to be treated as the *single source of truth*, some redundant data entry may be needed to keep the two systems in sync.
- One of the most common pitfalls with portfolio software is trying to get people to work with complex tools without sufficient training. Not only is training needed, it should also be archived for later use so people can refer to it when needed.
- Another strategy that works well is to have *local champions* who know the tool well and are accessible to the rank and file. These *local champions* are given sufficient training to enable them to function as subject matter experts for people who approach them.

CHAPTER SUMMARY

In this chapter, we explored the important role of portfolio information systems in creating a high-performing portfolio. We began by exploring why most portfolio management tools fail to meet expectations. We then went on to discuss the characteristics of a successful portfolio management system. First, we covered the intake process and its system needs, before moving on to the annual planning process and exploring its system needs. We then discussed the needs and the design for the portfolio transactions archival/retrieval system. We concluded with a listing of the general best practices for portfolio tools.

16

THE POLITICS OF PORTFOLIO MANAGEMENT

INTRODUCTION

There are few organizations, if any, that can claim to be apolitical. Political maneuvering is an everyday occurrence that we all deal with at the workplace. However, there is an extra dimension of politics to deal with when it comes to managing a portfolio successfully. The dynamics of portfolio management, with the attendant power to approve and terminate projects and control the spend associated with them, create a particularly fertile ground for political behaviors. In this chapter, we explore the following aspects of politics as it relates to portfolio management:

1. Explore what contributes to the political nature of portfolio management
2. Discuss the behaviors associated with the politics of avoidance
3. Discuss the behaviors associated with the politics of deflection
4. Describe a comprehensive list of effective strategies to combat political maneuvering
5. Explore staffing strategies that enable the portfolio office to navigate the political landscape to the portfolio's benefit

WHY IS PORTFOLIO MANAGEMENT SO POLITICAL?

Portfolio management seeks to control how the resources of the organization are spent. It selects the right efforts to work on, and once selected, it reports on how those efforts are progressing. It also prescribes corrective actions when things are not proceeding according to plan—including termination of efforts and possible transfer of ownership of troubled projects. Each of the areas mentioned earlier is quite political, and therefore, portfolio management, the sum total of these areas, is bound to be very political too.

Portfolio management, when done right, could make or mar the careers of people in the organization. Imagine an executive who is adept at managing their projects and delivering on time and on budget. Such a person would do very well with a functioning portfolio, because their projects would be clearly shown as performing better than their peers. Quite soon, such an executive could create a track record within the organization as a dependable performer who can be trusted to deliver on their projects and would be a likely candidate for mission-critical projects and soon, other responsibilities.

However, consider another, more commonly found scenario—an executive whose team isn't quite as well managed. The odds are that their projects would not be as successful. Such a team would show up in a less-than-flattering light when the portfolio is managed well. It would then be apparent to all that this team is not quite delivering on their project commitments and would be characterized by cost and schedule overruns, possibly even project terminations. In a typical organization, most projects and their owners are performing less than optimally. Consequently, very few executives are willing to have portfolio management cast the harsh spotlight of accountability on their projects.

Another area where portfolio management affects the political fortunes of executives is in the approval to start new projects. In an organization where there isn't a functioning portfolio, people start projects with little oversight. Often, no one even knows that a particular project has begun and is consuming resources until well after the fact. While everyone is off executing their pet projects, there may be little alignment between these projects and the stated strategic objectives of the organization. This would work quite differently when a functioning portfolio is in place, along with a robust intake management process. Only the projects aligned with a strategic road map would get approved, thus restricting the actions of executives who would like to spend resources on other (nonaligned) initiatives.

THE POLITICS OF AVOIDANCE

As seen in the previous section, few people prefer to be under the *performance microscope* that a portfolio would put them under. Therefore, the first instinct of project owners is to entirely avoid being part of the portfolio and all of the procedures that go with the portfolio. This gives rise to a set of avoidance behaviors, such as:

1. **Refusal to participate in standard phase gates**: This typically unfolds as a request for exemption from portfolio procedures, such as going

through the intake process. The most typical reason is that the project in question is *too critical* to be slowed down by going through the intake process.

2. **Refusal to use standard artifacts**: Project owners seeking to control the perception of their projects typically want to avoid the use of standard artifacts that may reveal the true status of their projects. This behavior manifests when the project owners insist on using their own nonstandard reporting templates, which may be vague or confusing. The typical reason given is often something like: "We don't want to waste time repurposing data from one format to another."

3. **Reluctance to be transparent**: Project owners who are practicing avoidance politics are reluctant to show the true state of the projects. The portfolio manager can expect them to resist (passively or actively) any portfolio procedure that would show the state of their projects; for example, modified earned value management. Along the same lines, they can also be expected to embrace (and refuse to discontinue) pseudo measures such as the red, yellow, and green system of reporting.

4. **Aversion to governance**: Finally, project owners wanting to avoid scrutiny are typically averse to governance. As explained in Chapter 17, governance should review and approve schedule changes, major scope changes, funding releases, and funding turnbacks. The political project owner would want to avoid all or most of this oversight and continue to work independent of governance.

POLITICS OF DEFLECTION

As explained in the previous section, the first instinct of project owners is to entirely avoid coming under the purview of portfolio management. But that cannot be a long-term strategy, for eventually the official policy would be to insist that everyone follow portfolio management guidelines. In that case, some project owners start adopting politics of deflection, such as:

1. **Complaining that the portfolio process is cumbersome**: This approach usually takes the form of blaming the new portfolio process for being unworkable—too many new steps, too many signoffs, and too much time taken to do it all. No matter how simple the process, the portfolio manager can expect an unfavorable comparison to the time when there was no process. "We'd like to follow the process, but it's simply unworkable," is the refrain adopted by the project team.

2. **Blaming the portfolio process for the slow pace of work**: This tactic tries to shift the blame to the portfolio process when the project fails to

deliver on time or on budget. "Much of our time was spent on portfolio process overhead instead of doing the project work," is what the project owners want to convey.

3. **Projecting that the portfolio staff is inefficient/inept**: This rather malicious approach tries to undermine the portfolio process by attacking the credibility of the portfolio staff. Examples include blaming portfolio staff for missing data, not providing data to the staff on time and then blaming them for incomplete presentation, and various other underhanded tactics. "How can the portfolio team criticize our project for nonperformance when they are doing so badly themselves," is the strategy being adopted here.

PLAYING POLITICS VERSUS BEING POLITICALLY CORRECT

Given all the complex political tactics adopted by various stakeholders, how can the portfolio manager drive change and succeed in creating a high-functioning portfolio? A broad approach is outlined in this section.

The portfolio manager needs to slowly create an environment where it is no longer politically acceptable to come in with an underperforming project—in other words, bring about a transformation where being an underperformer is a bad exception and untenable. This creates a self-regulating mechanism where politics work in favor of projects that are performing well rather than provide cover for bad actors. Here are the strategies for the portfolio manager to follow in effecting a change in the political climate as it regards portfolio management.

- **Strategy #1—Acquire the mandate for change**: Without a powerful and clear mandate, the portfolio manager will not make much headway. The portfolio initiative needs to be accompanied by a public, prominent mandate—preferably from the chief information officer (CIO). This would signal the other stakeholders that this effort is backed by leadership and cannot be evaded or dismissed easily. The portfolio manager would do well to hold a series of meetings with all of the stakeholders and start by invoking the mandate. It would also help to announce that regular status meetings will be held with the CIO and/or executive management regarding the progress of the rollout.

- **Strategy #2—Seek a powerful patron**: Playing politics is a skill and some people are just not cut out for it. A portfolio manager may be very skilled at portfolio techniques but could also be a total novice at politics. The portfolio initiative is too important to be left to the mercy of political opponents if the portfolio manager is not political enough. What is to be done in that case? One solution is to have the portfolio manager report to a powerful, seasoned executive on a day-to-day basis. This executive should have direction from the CIO to *protect* the portfolio manager and preserve the mission of rolling out portfolio management to the organization.

- **Strategy #3—Roll out changes gradually**: One of the mistakes that portfolio managers make in rolling out changes is to *throw the book* at their organization—namely, they try to do everything that they've read in the latest portfolio management tome. This almost never goes well for the following reasons:
  - The problems are deep-rooted and need a comprehensive fix
  - There is a finite capacity for change in an organization and trying to do too much will overwhelm the people and processes that are already in place
  - Not everyone is invested in making the changes—some of them would like to keep the current state (low or no portfolio management), as it benefits them

  Consequently, the *big bang* of change is met with a lot of resistance and may be used by some project owners to call into question the whole effort of portfolio management. The portfolio manager (and the initiative itself) may never recover from this body blow to their credibility. What's the remedy?

  A more prudent approach would be to introduce changes slowly, starting with the most significant, useful fixes. For example, it may be more important to have all the work come through an intake queue than have everyone follow the same template for monthly project performance reporting. Once everyone is using the intake queue, the portfolio manager should move on to the next pressing change that needs to be rolled out. In this approach, change can be assimilated as it occurs in more limited portions. Any complaints can be redressed quickly, or the latest change can even be canceled/rolled back as the portfolio team regroups to address the complaints. The biggest advantage of a system of *creeping change* is that project owners cannot find anything objectionable to protest against.

- **Strategy #4—Divide and conquer**: The portfolio manager should take all care to prevent a chorus of complaints arising from the organization. If a critical mass of people begin to complain about the changes made by the portfolio management process, it will be hard for even the sponsors of portfolio management (e.g., the CIO) to shield the portfolio manager from criticism. It may be wise to adopt a *divide and conquer* approach, as described in the next paragraph.

When process changes are introduced as part of the new portfolio management process, it is inevitable that many people will complain. There will definitely be some stakeholders who don't want any change at all—it's impossible to placate such parties. However, not all of the complaining parties may have the same grievances. Some stakeholders may have concerns that can be remediated without much effort. A diligent portfolio manager would do their best to accommodate as many parties as possible, at least in the beginning stages of the portfolio rollout. This may mean going above and beyond in helping the project owners with portfolio artifacts, if possible. It may mean being available for questions in addition to the standard training and documentation. It may also mean being flexible in matters of formats and templates. By doing all of the above (and whatever else is necessary), the portfolio manager can blunt the criticism generated by a few aggrieved parties who just want the whole portfolio process to go away.

- **Strategy #5—Knowing when to adopt pull versus push**: Consider a setup where the project owners have their own funding and the autonomy to spend it on projects that they deem appropriate. Now imagine a portfolio manager tasked with bringing all these project owners into the fold of portfolio management, with the attendant restrictions on their autonomy. It's very hard for a portfolio manager to make any inroads into changing the behaviors and outputs in such a situation. It can be expected that the project owners would fight at every turn to hold on to their autonomy and try to undermine the portfolio effort in every way possible. What can be done in such a situation? An alternative situation is outlined in the next paragraph.

  The previous approach can be characterized as a *push* situation, where the portfolio manager is attempting to *push* a new vision of how things should be run in the future—to an audience that is perfectly happy with how things are being run now. Consider a *pull* approach, where the project owners are compelled to come to the portfolio manager instead of the other way around. This can be accomplished by changing the funding model of the portfolio, as outlined in Chapter 4.

  A *pull* model fundamentally redefines the power vectors in the relationship between the portfolio and the project owners. In a *pull* model, the project owners are far more amenable to following portfolio procedures in the hopes of securing funding. Consequently the probability of conflict is greatly reduced under a *pull* model. There will still be some political wrangling, but the portfolio manager holds much more power and can direct the outcomes in ways that are in alignment with the portfolio's success.

Therefore, a portfolio manager should recommend actions to executive management that enable the functioning of a pull model rather than a push model.

- **Strategy #6—Drive for systemic fixes, not spot solutions**: Imagine a system where all project activity (labor and purchase orders) are billed to a project code in the financial system. Furthermore, the project code is auto-generated when the project comes through the intake process and a flag is set after governance review. That's an example of a systemic flow that makes no room for exceptions—every project has to come through the intake process, be reviewed by governance, and approved in the system before money can be spent against it.

  Now consider another system where the portfolio manager has to approve *outside the system* through an e-mail. Next, the project code is manually created and sent to Finance, who has to key it into the system before dollars can be spent against it. This is a recipe for confusion and an invitation for things to fall through the cracks, resulting in a backlash of criticism against the process. Wherever possible, the portfolio manager should focus on a *standard systemic flow* that addresses the needs of the vast majority of stakeholders. The flow need not be sophisticated or need expensive systems (consider SharePoint for low cost and rapid startup), but needs to be usable and to *work right* most of the time. Any other approach is vulnerable to failure and will attract needless negative attention.

- **Strategy #7—Practice strategic neutralization**: In many organizations, there is usually one person who plays a prominent role in opposing change. This person significantly holds up the rollout of the portfolio by opposing any and all process changes because it may dilute their long-held role or power. This dynamic is usually made harder because this person tends to be *indispensable* or a powerful stakeholder themselves. How can the portfolio manager deal with this roadblock that prevents any progress from being made?

  In this situation the portfolio manager needs to appeal directly to the CIO and point out how the needs of one individual are blocking the progress for the whole organization. In a frank and confidential manner, the portfolio manager needs to recommend to the CIO that the *blocker* be reassigned to a role that prevents them from opposing the portfolio. Little can be done if the CIO is not receptive to this course of action. It is this author's observation that portfolio efforts have been delayed for several years in organizations by a single obdurate individual, and things only began to move once that person left.

- **Strategy #8—Always protect the portfolio office**: The portfolio manager must always be aware of the political nature of their work and be hypervigilant against political foes seeking to damage the credibility of the portfolio process. It's important to realize that not everyone is aligned with establishing a high-functioning portfolio in the organization. Consequently, it's best to have everything in writing and it's always good to have portfolio decisions disseminated after governance meetings.

While the portfolio manager should endeavor to keep good relations with all stakeholders, the old adage of *trust but verify* would serve them well. In summary, the portfolio manager should be ready to provide proof or substantiation of all claims, especially as it relates to project performance and portfolio transactions.

- **Strategy #9—Anticipate problems and prepare to defuse**: Within a few cycles of the portfolio process, an astute portfolio manager can begin to predict the political actions of the stakeholders, especially ones that tend to be adversarial. As explained in a previous section, some political players seek to constantly evade and avoid while others try to deflect blame—the end goal is to defeat the portfolio system which affects their autonomy.

  In some time, the *habitual complainers* reveal themselves and tend to follow a predictable pattern of behavior. In such situations, the portfolio manager can preempt such behaviors by anticipating and taking advance action. For example, one common complaint following the rollout of a portfolio process is that *there was no training* or *no adequate training*. A smart move for a portfolio manager would be to attach a training guide, preferably with screenshots, to the e-mail announcing the process change. To make their position more formidable, they would also conduct multiple sessions of training and note who attended (this info can be useful to show, for example, that the complainers did not attend the training despite multiple sessions being made available to them—or even that they received the training and are still complaining about the lack of training).

- **Strategy #10—Cultivate goodwill at every opportunity**: The portfolio manager needs to grab every opportunity to sow goodwill in the organization. This includes all of the following suggestions and more:

  1. The portfolio team needs to always be helpful in navigating the portfolio management process and templates. Irrespective of the training offered, it is inevitable that people will still have questions about what to do and how to go about filling out the templates. At times like those, it's never a good idea to ask, "Why didn't you attend the training?" What will be appreciated is the portfolio staff being helpful in the moment of need. The portfolio manager needs to convey the expectation of

a service-oriented mindset to all of the portfolio staff—because ultimately it will reflect on the manager alone.

2. At least in the beginning of the portfolio rollout, there needs to be some flexibility in terms of templates and formats. It's important to choose battles wisely and almost never a good idea to be inflexible on the small things. This builds an image of the portfolio team being reasonable and easy to work with.

3. Simple is good as far as the project owners are concerned. It must be kept in mind that they have complex responsibilities in addition to following portfolio rules, and doing less in the portfolio is more from their perspective.

The cumulative result of cultivating goodwill is that the portfolio manager has a receptive audience while proposing additional process changes. It also creates a groundswell of support for the portfolio manager, which is useful when political foes want to attack the portfolio setup.

- **Strategy #11—Portfolio office should always super-communicate**: A portfolio manager can never communicate too much. At the same time, there are ways to over-communicate without being bothersome (such as spamming the organization with e-mails every hour). Consider the ways to super-communicate:

  1. All updates to portfolio process and templates need to be broadcast in multiple channels. For example, a process update announcement is provided in the SharePoint site along with the relevant templates and is followed up with an e-mail announcement, which also contains a link to the SharePoint site.

  2. It needs to be easy and straightforward for stakeholders to find things on the portfolio SharePoint site (see Chapter 20 for a detailed discussion about this topic).

  3. Wherever possible, the portfolio office needs to offer to configure automated alerts for stakeholders on the SharePoint site, enabling them to be immediately informed when something relevant changes on the site.

  4. It's very important to have a Frequently Asked Questions list on the SharePoint site that anticipates peoples' questions and redirects them appropriately.

Super-communication is a very smart political behavior to adopt, because it provides solid protection against the more serious complaint of: "The portfolio team never keeps us in the loop."

- **Strategy #12—Name and shame**: Some stakeholders are always trying to get an exception from following the process and having to do their part. Consider a project owner who is trying to wriggle out of creating a benefits statement for their project. In other words, they want to get the funding for a project without stating formally the benefits to be accrued by doing the project. Their end goal is to avoid providing a benefits statement that could then be held up as a comparison for what the project really provides by way of benefits at completion. Typically, they also want the portfolio manager to *provide cover* for them at the governance meeting by providing them with an exception. What should a portfolio manager do in such a situation?

The portfolio manager should stop *buffering* such bad actors from adverse visibility and let them assume accountability for their situation or position. As discussed, the goal of the portfolio manager should be to make it politically unviable for bad actors to continue flouting portfolio procedures.

In the example just quoted, the portfolio manager should ask the project owner in question to appear at the governance meeting and explain to the governance members why their ask is unaccompanied by the required artifacts (the portfolio manager should also advise the governance members in advance that this incomplete request is coming up for consideration and recommend not approving it).

As the portfolio management procedures take hold, *name and shame* becomes a very effective strategy to compel all actors to observe correct policies and procedures.

- **Strategy #13—Build on the momentum**: The road to establishing a portfolio is long and winding, but the portfolio manager eventually overcomes all hurdles and the portfolio starts gaining traction in terms of visibility and results. The journey is far from complete though, when all the potential improvements in the capabilities of the portfolio are considered. Therefore, the momentum needs to be maintained and preserved. What are the political behaviors to adopt in building the momentum?

  The progress of the portfolio journey can be highlighted in the following ways to create a sense of successful transformation and momentum:
  1. Celebrate successes of the portfolio, whether they be successful launches of projects, benefits delivered to the organization, or any noteworthy accomplishment
  2. Celebrate project owners/managers who achieve portfolio benchmarks—such as on-time, on-budget project completions
  3. Describe the road ahead in terms of future capabilities of the portfolio to create a sense of *better things to come*

The net effect of these activities is to create an impression of portfolio management delivering great value to the enterprise—this makes people want to align themselves with the portfolio team and comply with the policies and procedures.

THE EXPENDABLES

Rolling out a portfolio management process is quite a bruising political fight in most organizations. There are several players that are well entrenched in the current status quo and they can be expected to fight tooth and nail to oppose the new process that would directly impinge on their autonomy to select and run projects without oversight.

Few people would want to volunteer to step into the *hot seat* in those situations. Some executives and portfolio managers adopt a novel but expensive strategy to get change implemented while side-stepping the criticism and push-back that may come with it. That strategy is to hire a portfolio consultant who can be the *from-the-outside* face of all the tough decisions that need to be taken.

The consultant pushes through all of the painful iterations that need to be carried through and can finally leave, once the system is stable and well accepted. A side benefit is that the consultant can be summarily terminated as a sacrificial offering in one of the political maneuvers if need be. Although this sounds somewhat cruel and mercenary, this practice is actually fairly widespread and is part of the reason why portfolio consultants bill at such a high rate.

There is always a risk in hiring a portfolio consultant. For example, an organization could hire a generic *Big 4* consulting firm (as opposed to a specialized portfolio management expert) and the result would be that much time and money could be wasted in paper-pushing optics while no real work is done in advancing the portfolio's capabilities.

LEVELS OF PORTFOLIO CAPABILITY

Level 1

- Very political atmosphere, with spheres of influence carved out by respective players
- Portfolio has almost no influence on the respective players—almost no one follows portfolio policies and procedures
- Political considerations dominate in project selection and project performance readouts
- It's almost impossible to make progress on portfolio process improvements

Level 2

- Atmosphere still political, but there is an understanding that project performance supersedes politics

- Most stakeholders follow portfolio policies and procedures, with a few exceptions
- Strategic considerations dominate in project selection and performance, but some political jockeying exists
- Some resistance to portfolio process improvements

Level 3

- Political undercurrents in the atmosphere, but performance is paramount—there is a meritocracy of sorts
- All stakeholders follow portfolio policies and procedures; it is politically incorrect to be seen as not following portfolio policies and procedures
- Project selection and performance are driven by data and strategy considerations only
- Portfolio process improvements are received well, as no one wants to be seen as holding up progress

CHAPTER SUMMARY

In this chapter, we examined the reasons as to why portfolio management is so intensely political and how the real-life constraints of organizational politics affect the implementation of a portfolio. We covered at length the types of politics that are played by various stakeholders and what their end game is. We then covered a list of effective strategies practiced by successful portfolio managers to blunt the opposition of stakeholders who are opposed to the mission of the portfolio office. The chapter closes with a discussion of staffing strategies that could shield the portfolio office from political blowback.

17

PORTFOLIO GOVERNANCE

INTRODUCTION

In an ideal world, only the right projects would present themselves for funding, get funded quickly, and then proceed to deliver all of the promised benefits while executing on time and on budget. In the real world, almost none of those things happen. That's where portfolio governance comes in and tries to steer the actual course of events closer to what should ideally happen. This chapter explores the following aspects of governance as it relates to portfolio management:

1. Define governance and explain the role of governance in portfolio management
2. List the modes of operation of portfolio governance and distinguish between the modes
3. Explain the functions of governance under the routine portfolio operation mode
4. Describe the artifacts produced under the routine portfolio governance mode
5. Explain the functions of governance under the annual planning portfolio operation mode
6. Describe the artifacts produced under the annual planning portfolio governance mode
7. Describe the role of support systems in enabling portfolio governance to occur
8. Explore the ideal composition of portfolio governance
9. List the factors for portfolio governance success

WHAT IS PORTFOLIO GOVERNANCE?

Formally, governance is defined as the establishment of portfolio policies and continuous monitoring of their proper implementation by the members of the governing body of an organization. It includes the mechanisms required to balance the powers of the members (with the associated accountability). In the case of portfolio management, the primary duty of the governance body is to maintain and enhance the performance and viability of the portfolio.[1]

Portfolio governance is also used to refer to the team of senior leaders who carry out portfolio governance functions as previously defined. These leaders bear responsibility for the success of the portfolio and are empowered to make decisions about the portfolio, both tactical and strategic. They also have ownership of the direction of the portfolio.

TWO FUNDAMENTAL MODES OF PORTFOLIO GOVERNANCE

For discussion purposes, two modes of portfolio governance have been identified:

- Routine portfolio governance: The context for this mode of operation occurs during the whole year—focused on the current year.
- Portfolio governance during annual planning: The context for this mode of operation occurs during annual planning, which is basically planning for the next year. Annual planning is covered extensively in Chapter 3.

The details of these two modes are addressed separately in the next section.

ORCHESTRATION OF ROUTINE PORTFOLIO GOVERNANCE

Functions of Governance

The following functions are carried out by portfolio governance as part of the routine, round-the-year operation.

Function #1: Review of Current Portfolio Finances

The portfolio governance body convenes periodically to review the portfolio. One of the key functions during each session is to review the current state of the portfolio finances. Key assessments include the following:

- Is the portfolio spending as planned? Both underspend and overspend are undesirable.
- Has the portfolio allocated enough funds to projects? Again, a balance is needed here because either too much or too little will create a sub-optimal scenario.
- Does the portfolio look like it will end the year on budget or within a small variance? It is important for portfolio governance to become aware of how the year-end scenario is shaping up and start taking corrective actions.

Function #2: Approval of New Projects

A key activity that portfolio governance needs to perform is the approval and funding of new projects. In doing this activity, a delicate balance needs to be maintained: Appropriate proposals that are in line with the portfolio's mandate need to be approved as soon as possible, but questionable proposals need to be reviewed and blocked from proceeding. The key to achieving this balance lies in the application of project approval criteria as shown here:

- **Strategic fit**: How is this proposal a fit for the goals of this portfolio?
- **Cost**: What does this cost in terms of operating expenditure (OPEX) and capital expenditure (CAPEX)—by quarter, by year?
- **Strategic road map impact**: How does this project relate to our strategy? What capabilities will this project add to our strategic road map?
- **Benefits**: What is the return on investment (ROI)? Are the benefits soft (productivity savings, etc.) or hard (revenue, actual cost reduction in the general ledger, etc.)?
- **Resources needed**: What are the estimated resource requirements for this project?
- **Opportunity cost**: What is the cost of not doing this project? (This is good to have in case the proposal is not approved for funding.)

To enable portfolio governance to make the right decision, the fields listed above need to be populated and vetted for accuracy by the portfolio office before being presented to governance for a decision.

Function #3: Review of Project Performance

Portfolio performance ultimately depends on the performance of each project. Therefore, project performance is a key activity that needs to be undertaken by portfolio governance at every periodic session. Here are two factors that influence speedy and effective reviews of project performance:

- **Subjective performance data versus objective performance data**: Objective performance data is strongly preferred over subjective performance data for the purposes of project performance review by portfolio governance. Apart from the obvious advantages of making decisions based on objective data, there is the added advantage of objective data lending itself to sorting and ranking in a manner that allows portfolio governance to focus on the most essential projects in need of attention. Objective performance data allows portfolio governance to triage their efforts in an optimal manner.
- **Availability of modified earned value management (*mEVM*)**: mEVM has been extensively covered in this book as an ideal, low footprint methodology that delivers most or all of benefits of the objective performance measurement without the overhead. Accordingly, if mEVM is available and rolled out, it greatly enhances the value and effectiveness of portfolio governance's review. Having mEVM in place allows portfolio governance to view the entire portfolio at a glance (see the portfolio view in Chapter 11) and decide which projects seem to be in need of assistance to get back on track.

Function #4: Review of Enhanced Monitoring List

In Chapter 5, the concept of the enhanced monitoring list (EML) was introduced. The EML is a special list of underperforming projects that are closely monitored to help them recover. Accordingly, portfolio governance needs to review the EML at each regular session to decide one of the following three courses of action for each project on the EML:

- **Authorize removal of the project from the EML**: A project may be able to get back on track and already be performing as planned. In such cases, it is no longer necessary for the project to be featured on the EML.
- **Authorize continuation of the project on the EML**: A project may still be trying to get back on track and in the process of correcting its performance. In such cases, it is necessary for the project to remain on the EML.
- **Authorize termination of the project currently on the EML**: A project may be unable to improve its performance despite all efforts. Sometimes the issues are too deep rooted for the project to overcome and the portfolio needs to make a decision to terminate the project and redirect the resources elsewhere.

Function #5: Decision to Kill Projects

It is a fact of project management that some projects with serious performance problems cannot be remediated and have to be terminated. Other projects may be performing as planned but the anticipated benefits of the project may not be occurring, creating a need to terminate the project (see Function #6). Sometimes projects need to be terminated because of changing priorities or strategic direction. In all cases, portfolio governance has to make the decision to terminate a project after reviewing all the previous information. After portfolio governance makes the decision, the portfolio office records the decision and implements the same. It needs to be reiterated here that for any kind of performance-related action on a project, it is vital to have an objective performance measurement system—such as mEVM—that allows for data-driven decision making.

Function #6: Decision to Approve Change in Project Parameters

Projects are dynamic entities whose parameters cannot be predicted with certainty at the beginning of the project. Therefore, it's quite common for projects, even well-performing ones, to request an authorized change in scope, schedule, or budget parameters. It's a function of portfolio governance to approve these changes and the responsibility of the portfolio office to ensure that those changes are made part of the project's record. The portfolio office should ensure that any change to a project's parameters should also be reflected in the mEVM artifacts of the project—for example, the mEVM Excel template would change, causing a follow-up change in the graph and other artifacts as well.

Function #7: Review Benefit Delivery of Projects

The purpose of undertaking projects is to obtain the benefits promised by the projects. However, a significant proportion of projects fail to deliver the promised benefits. Hence, it is the duty of the portfolio governance body to monitor the delivery of benefits as promised and take corrective action if the benefits are not to be found. As covered in Chapter 7, there needs to be a portfolio benefits realization process that is managed by a portfolio benefits review council (PBRC). This council is the working body that goes through the details of the benefits of various projects and then presents the readout of these benefits (along with recommendations of courses of action) to portfolio governance. It's up to portfolio governance to decide whether to proceed with the recommendations. The actions that portfolio governance takes are recorded by the portfolio office and followed up on to ensure execution.

Meeting Frequency

How frequently does a portfolio governance body need to meet for the purpose of routine governance? Although it is a function of the combined availability of a group of senior people, the minimum recommended meeting frequency for routine governance is monthly. In the case of large portfolios, the monthly meeting may have to be long enough to accommodate all of the projects that have transactions at the session. An alternative is to hold more frequent sessions of smaller duration.

Artifacts of Routine Portfolio Governance

There are two important artifacts of the routine portfolio governance meeting:

1. The portfolio governance deck
2. The portfolio governance decision letter

These two artifacts are covered in detail in the next sections.

The Portfolio Governance Deck

The portfolio governance deck basically consists of the portfolio materials that are compiled by the portfolio office for review and decision by the portfolio governance body. Following are the five main components that should be in each meeting's deck.

Component #1: Portfolio Financial Performance Data

Portfolio financial performance data describes how the portfolio is doing from a financial perspective. Typically the following data is included while describing portfolio performance:

- **Overall portfolio budget**: This data shows how much money has been allocated to the portfolio for the current year. This needs to be broken out by OPEX and CAPEX.
- **Allocated funds versus remaining**: Of the total budget, how much has been allocated to various projects and how much is available for new projects?
- **Allocated versus released**: This data compares the amount of money allocated to various projects versus the amount of money that has actually been released.
- **Actual money spent by the portfolio**: This data shows what has actually been spent—the actuals—by the portfolio. Typically, there is a drill-down that should be available so governance can see the actuals by project.
- **Cost performance of the portfolio**: This data shows what the productivity is per dollar spent.
- **Schedule performance of the portfolio**: This data shows what the schedule progress is per dollar spent.

- **Estimate to completion versus allocated budget**: This data indicator attempts to estimate how much more money is needed for each project need before it becomes complete.

Component #2: New Project Proposals for Approval

New proposals that need approval and funding are included here for review and decision by portfolio governance. In order to facilitate a speedy decision, the following data should accompany each proposal, as pointed out in Chapter 2:

- **Strategic fit**: How is this proposal a fit for the goals of this portfolio?
- **Cost**: What does this cost in terms of OPEX and CAPEX—by quarter, by year?
- **Strategic road map impact**: How does this project relate to our strategy? What capabilities will this project add to our strategic road map?
- **Benefits**: What is the ROI? Are the benefits soft (productivity savings, etc.) or hard (revenue, actual cost reduction in the general ledger, etc.).
- **Resources needed**: What are the estimated resource requirements for this project?
- **Opportunity cost**: What is the cost of not doing this project? (This is good to have in case the proposal is not approved for funding.)

While only summary data regarding the proposals are needed for the portfolio governance review, the details substantiating the aforementioned information should be available in the Appendix (described under Component #5, later on).

Component #3: Project Performance Data

It's very important for portfolio governance to understand how each project is doing. At the same time, portfolio governance typically is composed of senior people whose time is limited. How can these opposing objectives be balanced? The answer lies in mEVM. With the help of mEVM, several aggregated portfolio views can be shown that display the macro picture with the ability to go to the detailed level if necessary. The most commonly used view, namely the monthly portfolio view, is shown in Table 17.1. The key elements of Table 17.1 are explained here:

- **Column A—Project Name**: This column contains the names of the projects in the portfolio. The name of each project is hyperlinked to the detailed mEVM data of that project.
- **Column B—Project Budget**: This column contains the official budget of each project in the portfolio.

- **Column C—How Much Money Did We Actually Spend?**: This column contains the actual money spent by each project. This is the same as the actual cost (AC) from the mEVM data.
- **Column D—How Much Work Should Have Been Completed?**: This column specifies how much work should have been done at this point in time in the project's timeline. This is the same as the planned value (PV) from the mEVM data.
- **Column E—How Much Work Did We Actually Complete?**: This column specifies the value of work that was actually completed. As described in Chapter 8, mEVM allots a specific amount of dollars for each deliverable. When those deliverables are complete, the project is allowed to *recognize* the value of the completed work.
- **Column F—What Is the Cost Efficiency (CE)?**: This column measures the ratio of Column E to Column C—that is, the ratio of completed work to the money being spent. CE is just another term for cost performance indicator (CPI) from the mEVM data.
- **Column G—What Is the Schedule Efficiency (SE)?**: This column measures the ratio of Column E to Column D—that is, the ratio of completed work to the work that was planned to complete (at this point in time). SE is just another term for schedule performance indicator (SPI) from the mEVM data.

The utility of this table lies in the fact that it shows the entire portfolio at a glance (Column A); along with the key data for each project (Columns B through G). Upon review, if portfolio governance wishes to know more about a certain project, the hyperlink embedded in the project's name jumps to a detailed snapshot of the project's mEVM data, allowing for a more detailed look at the project's performance. In this way, this table accomplishes the twin goals of summary top-level visibility and detailed line-level visibility.

Component #4: Enhanced Monitoring List

The EML has been introduced in prior chapters as a mechanism to spotlight troubled projects and help them recover or be terminated if they cannot be recovered. Table 17.2 shows what the EML would look like. The key elements of Table 17.2 are explained here:

- **Column A—Project Name**: This column contains the names of the projects in the portfolio. The name of each project is hyperlinked to the detailed mEVM data of that project.
- **Column B—Project Budget**: This column contains the official budget of each project in the portfolio.

Table 17.1 Monthly portfolio dashboard

Name of the Project	Total Budget of the Project	How Much Money Did We Actually Spend?	How Much Work Should Have Been Completed?	How Much Work Did We Actually Complete?	What Is the Cost Efficiency?	What Is the Schedule Efficiency?
(A)	(B)	(C)	(D)	(E)	(G)	(H)
Project 1	$500K	$100K	$100K	$75K	0.75	0.75
Project 2	$750K	$300K	$200K	$300K	1	1.5
Project 3	$100K	$50K	$40K	$25K	0.5	0.625
Project 4	$240K	$200K	$175K	$175K	0.875	1
Project 5	$1,000K	$500K	$450K	$500K	1	1.11
...	...	...	...	...	...	...
...	...	...	...	...	...	...

- **Column C—How Much Money Did We Actually Spend?**: This column contains the actual money spent by each project. This is the same as the AC from the mEVM data.
- **Column D—How Much Work Should Have Been Completed?**: This column specifies how much work should have been done at this point in time in the project's timeline. This is the same as the PV from the mEVM data.
- **Column E—How Much Work Did We Actually Complete?**: This column specifies the value of work that was actually completed. As described in Chapter 8, mEVM allots a specific amount of dollars for each deliverable. When those deliverables are complete, the project is allowed to *recognize* the value of the completed work.
- **Column F—What Is the CE?**: This column measures the ratio of Column E to Column C—that is, the ratio of completed work to the money being spent.
- **Column G—What Is the SE?**: This column measures the ratio of Column E to Column D—that is, the ratio of completed work to the work that was planned to complete (at this point in time).
- **Column H—Reason for Being Listed on the EML**: This column states the reason why this project has been placed on the EML. The three main reasons for a project to be on the EML are:
  1. Nonperformance on cost—also known as Low CE
  2. Nonperformance on schedule—also known as Low SE
  3. Nondelivery of benefits—also known as NDB

It needs to be noted that a project could suffer from more than one type of deficiency, as shown in Table 17.2.

Component #5: Appendix with Raw Data

It needs to be remembered that all of the data shown in the first four components of the portfolio deck are summary data that are meant for quick review by the portfolio governance members. However, the members may want to dive deep into certain items to understand the details before coming to a decision. For completeness, all the raw data (from which the summary data is derived) is stored in the appendix. The raw data consists of the following:

- Details underlying portfolio actuals (for Component #1)
- Details underlying new project proposals (for Component #2)
- Details underlying project performance data (for Component #3)
- Details underlying project performance data for the projects on the EML (for Component #4)

Table 17.2 Enhanced monitoring list

Name of the Project (A)	Total Budget of the Project (B)	How Much Money Did We Actually Spend? (C)	How Much Work Should Have Been Completed? (D)	How Much Work Did We Actually Complete? (E)	What Is the Cost Efficiency? (F)	What Is the Schedule Efficiency? (G)	Reason for Being Listed on the EML (H)	Trend History on the EML (I)
Project A	$500K	$600K	$500K	$365K	0.61	0.73	Low CE	Improving
Project B	$750K	$600K	$600K	$300K	0.50	0.5	Low CE, SE	No change
Project C	$100K	$150K	$100K	$75K	0.50	0.75	Low CE	Deterioration
Project D	$240K	$200K	$175K	$175K	0.88	1	NDB	Improving
Project E	$1,000K	$500K	$500K	$300K	0.60	0.6	Low CE, SE	Improving
Project F	$1,000K	$700K	$600K	$700K	1.00	1.16	NDB	No change
...	...	...	...	...	...	...		
...	...	...	...	...	...	...		

To make navigation easy, it is recommended that hyperlinks be embedded in the main body of the deck which when clicked upon would jump to the relevant section in the appendix.

The Portfolio Governance Decision Letter

Each time portfolio governance meets, there are many decisions to be made. Potentially every item listed on the portfolio governance deck could result in a decision. Therefore, there is a need to create a format where all of these decisions can be recorded and disseminated. This is where the portfolio governance decision letter comes in. This is a document that contains all the decisions that were made by portfolio governance at a particular session. This document is then disseminated within the organization to ensure visibility to the decisions contained therein.

Relation to Other Governance Bodies

As mentioned before, portfolio governance is composed of senior people whose time is at a premium. One way of mitigating this executive availability constraint is to divide some of the governance load among other parties. This arrangement will typically take the following forms:

- **The portfolio subcommittee**: This is a subordinate body to portfolio governance and has been granted oversight authority within limits. For example, the subcommittee could approve new project proposals up to a nominal amount of $100K. The subcommittee could also review other components of the portfolio governance deck and make recommendations for portfolio governance's consideration. For example, the subcommittee could review the performance details of projects and recommend that some of them be placed on the EML. Alternatively, they could recognize the improving trend of certain projects already on the EML and recommend that these be taken off the EML. The advantage of this arrangement is that the subcommittee can spare the time to go through the details and then make appropriate recommendations for final approval by the main body (portfolio governance). To make this arrangement functional, there needs to be a trusted working relationship between the main body and the subcommittee.
- **The benefit review board**: One of the big challenges in portfolio management is ensuring that projects deliver on their promised benefits. As covered extensively in Chapter 7, benefits realization management practices need to be in place to ensure that benefits are delivered as promised. One of those benefit management practices is the regular functioning of

a PBRC. The PBRC reviews benefits in detail and then makes recommendations to the portfolio governance body. For example, consider a project that is not delivering benefits as planned. The PBRC reviews this project as part of its review of all benefit-bearing projects and makes a recommendation whether to terminate the project or not. This recommendation is supplied to the portfolio governance body, which then has to make the final decision to terminate the project.

ORCHESTRATION OF ANNUAL PLANNING PORTFOLIO GOVERNANCE

Functions of Governance

Annual planning is the exercise to comprehensively analyze the organization's demand for the next year and decide which proposals to fund. The following functions are carried out by portfolio governance as part of the annual planning activity.

Function #1: Authorization of the Annual Planning Exercise

The portfolio governance body authorizes the portfolio office to kick off the annual planning exercise. Before socializing changes to templates and processes with the rest of the organization, the portfolio office first previews the same with the governance body for their approval. After portfolio governance provides their approval, the portfolio office proceeds to launch annual planning activities—such as the planning workshop—where the new templates and process changes are socialized.

Function #2: Validate the Strategic Road Map and Strategic Priorities for the Coming Year

Before the annual planning year can start, portfolio governance needs to review and validate the strategic road map. Figure 17.1 shows the strategic multi-year road map for Dimension A. Similar to this road map for Dimension A, there are other road maps for the other strategic dimensions. Prior to embarking on gathering the demand data from the organization, the following is required from portfolio governance:

- Confirmation that the set of strategic dimensions is valid for the annual planning activity
- Confirmation that all projects need to align themselves to one or more strategic dimensions to be included in the annual planning exercise

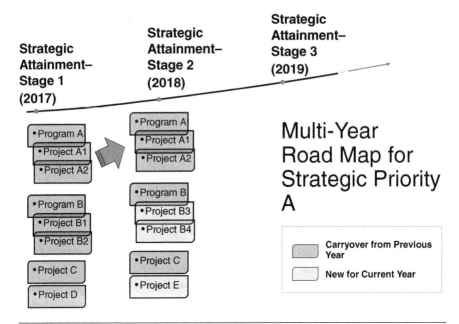

Figure 17.1   Strategic multi-year road map for Dimension A

This function is very important because the whole exercise of annual planning needs to be in alignment with the strategic road map of the organization. By confirming the strategic priorities ahead of the annual planning effort, portfolio governance sets the stage for the rest of the exercise.

Function #3: Review Aggregated Annual Planning Data

The first step in annual planning involves the collection of demand from all the stakeholders in the organization. The second step is to compile this demand data into an aggregated form for easy analysis. It's the portfolio governance's role to review this data and make key decisions on whether a certain project is above the line or below the line. One of the key artifacts for portfolio governance to review is the bubble chart shown in Figure 17.2. This bubble chart shows how all the demand gathered during annual planning is distributed from a cost/risk/benefit perspective.

Review of the bubble chart is necessary to decide which projects offer the best cost/benefit/strategic value ratio and factor that decision into the final *above-the-line/below-the-line* determination (for more details please see Chapter 3 on annual planning).

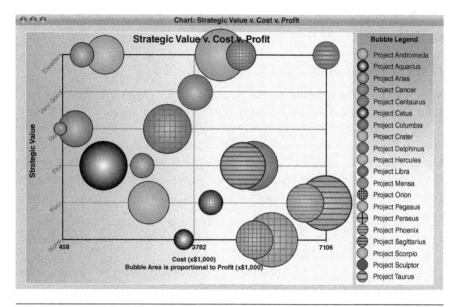

Figure 17.2   Bubble chart showing the distribution of annual planning demand

It also falls to portfolio governance to review how the mass of collected demand aligns to various strategic dimensions—for example, certain priorities could be oversubscribed while others may not have sufficient projects that identify as belonging to those undersubscribed dimensions. In that case, portfolio management has two decisions to make:

1. Decide which of the projects aligned to the oversubscribed dimensions should get funding. For example, if 15 projects are aligned to Strategic Dimension A, portfolio governance should decide if all 15 would get funded (leaving less money for other strategic dimensions) or if only some of the 15 should be funded—and if so, which ones.

2. Decide what should be done to mitigate the undersubscribed dimensions. For example, if only two projects are aligning themselves to Strategic Dimension B, it may not be enough to achieve the required transformation in that dimension's multi-year journey. Portfolio governance needs to decide if they will direct certain project owners to make changes in their submission or make new submissions that mitigate the gaps.

Finally, portfolio governance may need to review the templates associated with each project to get more in-depth information about each project, as necessary. After reviewing all of the oversubscribed and undersubscribed dimensions, portfolio governance needs to decide on the ranking of the projects. This ranking information is provided to the portfolio office, which then publishes a ranked list of projects comprising the annual planning demand.

Function #4: Supply Demand Matching

Most organizations start the annual planning exercise with a preliminary budget number that represents the total amount of money available to fund projects in the new year. The challenge then is to maximize the value delivered to the organization within the constraints of this budget number. When it is found that the aggregate demand exceeds the supply of funds, it is then a function for portfolio governance to determine what needs to be done to bridge the gap between supply and demand, which is addressed as follows:

- Portfolio governance first reviews the available funds along with the demand of the top-ranked projects and determines what's *above-the-line* (funded) versus *below-the-line* (unfunded).
- The implications of not funding certain projects (also known as the *cost of not doing*) is discussed among the portfolio governance body and shared with the organization and other stakeholders as appropriate. The aim is to determine if the present *above/below* line proposal creates unacceptable risk to the organization.
- Typically, some relief is obtained from Finance in the form of additional funds, which makes it possible to squeeze in a few more projects above the line. Portfolio governance will need to make the decision about which projects should get these funds and get *above the line.*
- Alternatively, if no additional funds become available, portfolio governance may decide that funds need to be carved out of existing *above-the-line* projects to provide for a few *below-the-line* projects. Portfolio governance also needs to decide which projects need to pare down their demand and by how much.
- After some jostling and rearranging, a final configuration emerges that portfolio governance needs to make official as the annual planning list. In the context of this final list, the portfolio office updates the strategic road maps (Figure 17.1) for all the dimensions to show *funded versus nonfunded* projects so that portfolio governance fully understands the impact of the funding decisions to the strategic road map.
- The *below-the-line* projects are turned into a queue that can absorb any new funds that may become available during the year. Portfolio governance reviews and approves the *queue* and confirms that these projects will be first in line for any new funding that becomes available.

Meeting Frequency

How frequently does a portfolio governance body need to meet for the purpose of annual planning governance? Although the name *annual planning* denotes once a year, in reality, the annual planning exercise takes several months to successfully accomplish. Therefore, it is appropriate to plan for several sessions of portfolio governance to transact the necessary actions for annual planning leadership. Since the earlier constraint of availability of a group of senior people still applies, some organizations prefer to hold a few combined sessions where both the annual planning activity and the routine governance activity occur. Given the difficulty of holding long sessions with the limited availability of executive members, one mitigation strategy followed by different organizations is to circulate annual planning materials for offline review by the portfolio governance members. Only the final decisions need to be included in the live session.

Meeting Artifacts

Before we explain the list of meeting artifacts involved in the annual planning governance, it needs to be remembered that these artifacts are created and reviewed in a progression of time and are not all present at one meeting. (This is in marked contrast to routine portfolio governance, where the governance deck and the decision letter are featured at every session.) In other words, the first annual planning meeting may involve just a review of the annual planning materials package, which consists of the unranked list of projects, multi-year strategy road map, and the bubble chart. Once portfolio governance has had a chance to review these artifacts and articulate a preliminary priority, the portfolio office is able to create and table the next artifact (preliminary ranked list of projects) for discussion at the next session, and so on until the final annual planning list is created. Here is the combined list of the artifacts used as part of the annual planning exercise:

1. Unranked list of projects
2. Multi-year strategy road map
3. Bubble chart
4. Preliminary ranked list of projects
5. Above-the-line/Below-the-line determination list
6. Final annual planning ranked list of projects

Relation to Other Governance Bodies

Annual planning is characterized by the need to analyze huge amounts of demand data in order to arrive at the optimal mix of projects to fund for the next year. Although the portfolio office distills the mass of data into a few concise artifacts—such as the annual planning materials package—it is still a time-intensive task that the senior executives who comprise the portfolio governance body are confronted with. One way of dealing with this situation is to divide some of the governance load among other parties. This arrangement typically takes the following forms:

- **The portfolio subcommittee:** This is the same subordinate body that was seen earlier as part of the routine portfolio governance. The subcommittee works under the authority of portfolio governance and has been granted oversight authority within limits. The subcommittee would go through the annual planning demand data and make requests for changes or clarifications. It may challenge some of the data that does not stand up to scrutiny. It may even come up with a suggested ranking, which would then be approved (with modifications, if necessary) by the portfolio governance body. The advantage of this arrangement is that the subcommittee can spare the time to sift through the mass of data generated by annual planning demand gathering and ensure that all of the data is correct and dependable for portfolio governance to make decisions. An added advantage is that the subcommittee can make appropriate recommendations for final approval by the main body (portfolio governance). To make this arrangement functional, there needs to be a trusted working relationship between the main body and the subcommittee.

- **The strategic priority advocates:** The role of the priority advocate was introduced in Chapter 3. The priority advocate is someone who is the steward of a particular strategic priority. The roles and responsibility of the priority advocate are described as follows:
  - The priority advocate owns a portion of the strategic road map and has been designated as the point person for that strategic priority of the organization's strategic road map. (Refer to Chapter 2 to review how the different priorities fit into the overall strategic road map.)
  - The priority advocate engages with the project owners and analyzes how/whether their project fits with that particular strategic priority (hence the term *priority advocate*).
  - The priority advocate validates the strategic impact of a project proposal as documented in the annual planning template and assigns a relative score, if necessary. (Example: Project A1 could have a relative score of nine, whereas Project C could have a score of seven as it relates to influencing the strategic priority. Alternatively, strategic impact (to the same specific priority) can be described as high/medium/low).

- It's worth noting here that there may be several priority advocates, each responsible for a particular priority in the strategic road map.

Portfolio governance relies heavily on the priority advocate to ensure that the whole annual planning exercise is strategically meaningful. The priority advocate ensures that projects are really aligned to the strategic priorities and provides portfolio governance with the confidence to proceed on that basis to make the right decisions. Finally, after the conclusion of the annual planning exercise, the priority advocate updates the strategic road map with new projects that fit within the priority and get funded as part of the annual planning.

ENABLING GOVERNANCE: SUPPORT SYSTEMS

For governance to work well, it needs to be bolstered by support systems that enable governance to provide effective decision making. Here are the key support systems that are necessary for optimal governance function:

1. **mEVM**: An objective system of measurement such as mEVM is vital to focusing governance time on the projects that really need direction from governance. Without mEVM, governance would have to conduct a detailed review of all projects—for which there is simply not enough time.
2. **Historical portfolio repository (HPR)**: HPR is a tool that preserves all project transactions and decisions by portfolio governance and retrieves them for later review. This is a very useful device for portfolio governance, who may need to review past decisions and any follow-ups indicated in the past for certain projects. It may also help with decision making with projects.
3. **Strategic map**: Having a validated, official strategic road map immensely helps portfolio governance in making decisions relating to project intake and funding. The presence of this artifact is very important because the whole exercise of annual planning is dependent on aligning projects with the strategic road map of the organization. By confirming the strategic priorities ahead of the annual planning effort, portfolio governance sets the stage for the rest of the exercise. Without a strategic map, portfolio governance would have to proceed on a *best guess*.
4. **Strategic earned value (SEV)**: The concept of SEV was extensively explored in Chapter 11. Basically, SEV allows the measurement of whether a project succeeded in delivering the strategic value it promised. This is a very useful technique for portfolio governance because delivery of strategic value should drive most of the decision making in a portfolio.

EXPLORING GOVERNANCE COMPOSITION

Is governance essentially a case of managing by committee? Traditionally, a committee structure of leadership is associated with deflection of responsibility and lack of accountability. One could make a case that it would be more effective for the portfolio to be run under a single point of leadership—which does happen in many places. However, the corporate trend points toward *participatory democracy*—a group of leaders getting together to make decisions that work for all of them, hold them accountable, and still achieve the desired impact on the ground. Therefore, the predominant structure for portfolio governance tends to be a team of people.

Another reason governance composition needs to be more than one person is to ensure representation of all of the stakeholders' interests. An information technology (IT) portfolio might have its governance body consist of all of the chief information officer's directs, ensuring that the portfolio works for all of IT and not just engineering, for example. It's also a good idea to include business representation in the portfolio governance body because the ultimate purpose of a portfolio is to serve the interests of the business.

SUCCESS FACTORS FOR GOVERNANCE

Governance is successful when all (or most) of the following concepts hold true.

#1 Governance Members Have a (Significant) Stake in the Portfolio's Functioning

Sometimes organizations will have governance members who are not directly related to the portfolio's main body of work. Such people, while well meaning, are simply not driven enough to demand optimal output from the portfolio. They are also not very insistent on taking the portfolio to the next level. Their lukewarm interest in the portfolio's capabilities and throughput translates into a suboptimal functioning for the portfolio.

#2 Governance Members Have the Mandate to Make Hard Decisions

As the saying goes, the buck stops with the governance as far as the portfolio is concerned. The portfolio governance members need to be able to make hard decisions, such as:

- Saying *no* to bad projects—not even starting them
- Terminating bad projects that are beginning to underperform

- Validating projects if they are indeed delivering benefits as promised
- Rebalancing the portfolio as needed

#3 Governance Is Empowered with Tools and Data

For governance to work, they need to be provided with the right tools and data about project performance so they can make the right decisions. That's precisely why there is a strong need for a system like mEVM to provide governance with clear, objective data. In the absence of such a system, even a motivated and empowered governance team is prevented from making the right decisions.

#4 Governance Members Are Coachable in Portfolio Theory

Governance members must recognize that portfolio management is a science and that there is opportunity for their portfolio to evolve in terms of capability. Consequently, they should be coachable/receptive in understanding how to function in a way that enables the portfolio to get to the next level. I have found that portfolios work best when governance understands how to integrate new tools or new techniques to aid their decision making.

CAUTIONARY NOTE ABOUT GOVERNANCE

Having a governance body in place is a great thing for a portfolio, as long as the governance body actively monitors the portfolio and makes the right decisions. At the same time, governance bodies would do well to guard against becoming the *masters of the tea ceremony*. Imagine a portfolio with regularly scheduled governance meetings, with all the trappings that go with it—that is, well-prepared meeting materials, followed by detailed meeting minutes and documented decision letters. Does this always result in a well-functioning portfolio? No, it doesn't.

An indicator of the portfolio's success over the years can be measured using one of the mEVM aggregation measures introduced in Chapter 10. Figure 17.3 shows the multi-year portfolio graph, which provides visibility into the portfolio's performance over the course of a multi-year journey. A high-performing portfolio should show an increasing trend in CE and SE over the years. This increasing trend is only possible when the portfolio governance is willing to make the hard decisions of terminating bad projects, funding the right projects, and being an effective steward of portfolio funds and output.

Surprisingly, many portfolios are reluctant to tackle the politically harder task of actual governance and instead settle for observing the outward

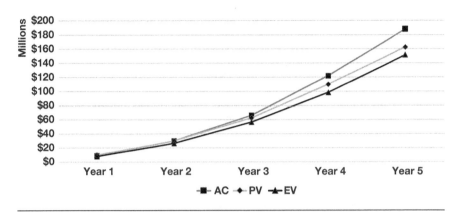

Figure 17.3   Graph showing multi-year performance of the portfolio

appearances—much like an elegant tea ceremony that transacts no real business. In this kind of situation, the only way to ensure that governance is effective comes from executive vision at the top.

LEVELS OF PORTFOLIO CAPABILITY

Level 1

- The organization has a limited understanding of the importance of governance. Consequently, there is little governance overseeing the portfolio.
- Due to the limited footprint of governance, it may not be possible to draw a distinction between the different modes of governance (such as routine mode and annual planning mode).
- If governance exists, its role may be limited to a cursory review of project performance as opposed to a comprehensive review of portfolio health indicators.
- There may not be defined artifacts generated by routine portfolio governance.
- The concept of the EML may not be known or put into practice. This hampers the function of governance.
- There may not be other defined bodies—such as the benefit review board—that can complement the function of governance. Consequently, governance may have to perform those functions or do without them.
- Governance support systems such as mEVM, the HPR, the strategic road map, and SEV may not exist at all or may function at a very limited level. Therefore, the governance has very little by way of meaningful tools to work with.

- Few or none of the success factors of governance are found.
- At this level, governance is symbolic and not really delivering meaningful direction or value to the portfolio.

Level 2

- The organization has a good understanding of the importance of governance. Consequently, there is governance oversight over all the major areas of the portfolio.
- There is a clear distinction between the different modes of governance (such as routine mode and annual planning mode).
- Governance's function covers a review of project performance as well as a comprehensive review of portfolio health indicators.
- There is a clear definition of artifacts generated by routine portfolio governance as well as annual planning portfolio governance. These artifacts are widely used and circulated.
- The concept of the EML may be known, but the organization has not implemented an objective system of measurement such as mEVM to support the EML.
- The function of governance is complemented by related bodies such as the benefit review board and a portfolio subcommittee.
- Some governance support systems out of the four important ones (such as mEVM, the HPR, the strategic road map, and SEV) may exist. The one important system that is not likely to exist is mEVM, thus depriving the governance of objective data.
- Some success factors of governance are present, helping the governance body to function better. Governance may still not be able to terminate underperforming projects in a prompt manner.
- At this level, governance is present and serving a useful function to the portfolio, although much more optimization is possible.

Level 3

- The organization has a highly evolved understanding of the importance of governance. Consequently, there is governance oversight over all of the major areas of the portfolio.
- There is a clear distinction between the different modes of governance (such as routine mode and annual planning mode).
- Governance's function is comprehensive and covers both a review of project performance as well as a detailed review of portfolio health indicators.

- There is a clear definition of artifacts generated by routine portfolio governance as well as annual planning portfolio governance. These artifacts are widely used and circulated.
- An EML is in place along with well-defined entry and exit criteria. The functioning of the EML is aided by the presence of a well-implemented mEVM framework.
- The function of governance is complemented by related bodies such as the benefit review board and portfolio subcommittee.
- All the key governance support systems (such as mEVM, the HPR, the strategic road map, and SEV) exist and are functioning soundly. The cornerstone of objective measurement, namely mEVM, is present and thriving.
- All success factors of governance are present, creating an optimal environment for portfolio governance to function. Portfolio governance is able to perform the difficult task of terminating underperforming projects in a timely manner.
- At this level, governance is not only present but also optimized. It effectively performs its primary role of providing direction and oversight, thus creating the conditions for a high-performing portfolio.

CHAPTER SUMMARY

In this chapter, we explored the vital role of governance in creating a high-performing portfolio. First, we defined portfolio governance and then listed the modes of operation of governance. We also covered at length the typical functions of governance under the routine portfolio operation mode as well as the annual planning mode. Next, we explored in detail the artifacts created as part of the two modes of operation. We went on to describe the role of governance support systems, which are crucial for the success of portfolio governance. We also covered the ideal composition of portfolio governance and explored the pros and cons of committee structure. We then looked at the success factors that enable portfolio governance to succeed in carrying out its mission of providing direction to the portfolio. The chapter expresses a cautionary note that advises how to ensure that a portfolio does not devolve into a ceremonial body, then concludes with a listing of the levels of maturity in this key portfolio capability.

NOTE

1. *Definition of Governance,* http://www.businessdictionary.com/definition/governance.html

Part IV

The Support Systems that Decide Success

18

ROLE OF THE CHIEF
INFORMATION OFFICER

INTRODUCTION

More than ever, the office of the chief information officer (CIO) and the portfolio office are connected in a symbiotic relationship in modern organizations. This chapter will cover the following topics to illustrate this two-way dependence:

1. Explore how the CIO's mandate has a direct impact on the portfolio
2. Demonstrate how the CIO is the biggest beneficiary of a high-functioning portfolio
3. Demonstrate how the CIO is also the biggest enabler for a high-functioning portfolio office
4. Enumerate the services the CIO can expect from the portfolio in addition to a list of actions the CIO can take to empower the portfolio

A CIO'S MANDATE

A CIO's mandate is essentially that of a change agent. The CIO is expected to effect a change for the better in systems, performance, and capabilities of the organization. However, this mandate is not without constraints—the biggest one being a time horizon within which the CIO is expected to move the needle or move out of the role. The time-tested way to effect change is through a well-managed portfolio of projects—a project being a temporary endeavor that results in an end product/capability/service.

Therefore, it is very essential for a CIO to have or create a high-functioning portfolio. Without such a mechanism in place, the bad projects will crowd out the good, slowing down enterprise transformation to a crawl and ultimately leading to a loss of confidence in the CIO's leadership. To avoid this negative spiral, the CIO needs to put every possible resource into supporting the portfolio office.

HOW THE CIO IS THE BIGGEST BENEFICIARY OF THE PORTFOLIO OFFICE

Benefit #1: Portfolio Office Can Help Define the Strategic Transformation Journey in Concrete Terms (and Call Out When that Is Not Done)

Although every CIO starts with a lofty strategic vision, there must be a concrete description of what is sought to be achieved and how to get there. That document is the strategic road map and that, in turn, can be decomposed into more tactical units of achievement. The portfolio office relies heavily on an accepted strategic road map to carry out much of its function. Conversely, the portfolio office is among the first entities to notice and call out (politics permitting) the absence of a strategic road map. This in itself is a big help for the CIO—the realization that there is not a credible strategic road map that forms the basis of the CIO's strategic transformation vision. Although it is not the primary function of the portfolio office to come up with a strategic road map, the portfolio office can be a very effective partner in the effort. Furthermore, the portfolio office can ensure that the strategic road map is usable and grounded in reality, thus avoiding the usual trap of the strategy becoming *credenzaware*.

Benefit #2: Portfolio Office Can Help Choose the Right Projects (and Block the Wrong Ones)

Once a strategic road map is created and agreed to by all stakeholders, it forms the underpinning for all projects being considered for execution. In other words, only those projects need to be taken up that are in accordance with the road map and will take the vision forward. The portfolio office is the gatekeeper that manages the intake process and ensures that incoming projects are vetted before execution. Without this function of the portfolio office, the CIO is likely to be blindsided by projects that waste time and money while contributing little to the strategic agenda. The best time to stop a bad project is at the very beginning, which is what the portfolio office does through intake management.

Benefit #3: Portfolio Office Can Help Decide Whether to Terminate Underperforming Legacy Projects

Most CIOs have to contend with legacy projects that they inherited from previous management. Some of these projects are performing as planned and are not a problem. Often, however, these projects are long-running debacles and may have played a role in the previous CIO's departure. Once inherited, the tough decisions remain as to whether this project should be kept alive in order to avoid writing off the millions already spent in the years past—or whether there is a risk of throwing good money after bad in trying to salvage these projects. Here is where a competent portfolio office can save the day and provide the right recommendation to the CIO. First of all, the portfolio office would get past the simplistic analysis of *sunk cost versus cut losses* and proceed in the following way:

- Does the legacy project still fit the new strategic road map as currently agreed by all stakeholders?
- What has been the strategic attainment of the legacy project until this point? How does this compare to the promised strategic attainment?
- If the project has been running for a while, has it already delivered some benefit? How do the delivered benefits compare with the promised benefit statement?

If the project clears all of the above hurdles, the portfolio team would also define a modified earned value management (*mEVM*) plan for the project that determines the following:

- Is the project performing to plan while staying on track with budget and schedule?
- Is the trend worsening or getting better?
- Extrapolating to the future, what will the project cost when it is finally done?

With all of the previously mentioned data and analysis, the portfolio office can deliver a detailed, data-supported recommendation about keeping legacy projects and prevent the CIO from the twin dangers of:

1. Terminating a viable project
2. Not terminating an unviable debacle

Benefit #4: The Portfolio Office Can Call the CIO's Attention to Trouble before It's Too Late

The bane of most CIOs is that people who report to them do not bring them bad news until it's almost too late to do anything. In the defense of people who report to the CIO, many of them are not sure about the true status of projects and the transformation effort themselves. As covered extensively in Chapter 5, there are a whole host of factors that prevent the true status of projects from being visible to stakeholders (including the CIO, who is ultimately accountable for the success of the strategic transformation).

This is where the portfolio office offers an early warning that may save the larger transformation effort. By deploying mEVM and using a Pareto approach, the portfolio focuses on the few projects that are in need of executive intervention. Not only does this focus executive attention where it is truly needed, it also does so in a timely manner, where the executive still has room to maneuver (for example, replacing the current project manager on a troubled project with a more seasoned veteran might correct the project's trajectory before time and money are lost). By taking quick, effective action on the few projects that need it, the overall transformation effort is kept on track. The portfolio office enables this corrective feedback loop by accurately focusing on the trailing projects.

Benefit #5: The Portfolio Office Creates a Project Meritocracy that Supersedes the Politics

Every organization has its own kind of politics and the CIO has to find a way to execute the strategic transformation agenda within that political framework. As detailed in Chapter 16, the political behaviors adopted by various players (in most organizations) serve to either obscure the visibility of true project performance or evade oversight by governance. Prevailing rules of political correctness allow subpar execution to be hidden until the damage is obvious to all and beyond recovery.

However, the portfolio office, when supported by the CIO, can create an environment where the true performance of projects is visible to all. Through mEVM and other tools, the portfolio office creates transparency, which puts underperformers on the defensive and accelerates either recovery or termination. Over time (within a year, in most cases), there is a sort of meritocracy in place in which underperformers find it politically unviable to operate for long. This self-regulating mechanism is a huge catalyst to the eventual achievement of the strategic transformation effort—underperformers will either actively try to improve or bow out.

Benefit #6: The Portfolio Office Can Give the Governance Body a Framework to Operate Within

In organizations that are advanced enough to have knowledgeable and empowered governance bodies, there remains a final hurdle—the governance members do not have actionable data. What is actionable data? It's objective, quantitative data that enables decision making with a high degree of confidence. Instead, governance is often fed reams of paper that hold pseudo tracking information—such as the red, yellow, and green system of reporting—or extensive, often subjective, verbiage describing status. Such information wastes the time of governance and camouflages the true problem projects. Therefore, although the CIO looks to the governance body to provide effective oversight of the portfolio, the governance members are unable to deliver on that expectation for lack of actionable data.

The portfolio office can remedy this situation through the mEVM system and its readouts. In the place of pseudo performance data that means little, mEVM provides objective information about project performance in the form of an easy graph. It also provides drill-down capability and trend information, as well as the basis for assessing whether or not turnaround efforts are working for underperforming projects. Taken together, this provides the governance body with a solid basis to make sound decisions, and furthermore, enables them to govern by exception—in other words, spend time on the projects that need it.

Benefit #7: The Portfolio Office Can Orchestrate Annual Strategic Planning for the CIO

Without annual planning, the organization is at a distinct disadvantage in optimally matching supply and demand. Furthermore, as covered in Chapter 3, annual planning eliminates redundancy and promotes much needed conversation about program alignment, prioritization, and dependency. The various artifacts created during annual planning help the organization make the right decisions about aligning demand with the strategic road map. In short, annual planning is a vital prerequisite to enabling the strategic transformation effort of the CIO.

The portfolio office is the body that owns and delivers annual planning. It works through the templates, the data gathering rounds, and orchestrates the multiple passes involved in the planning activity and finally provides the analyzed information to the decision makers (including the CIO) for the right projects to pick for the year. This exercise vastly clarifies the picture of what the organization seeks to achieve and how that picture is directly related to the strategic road map. Furthermore, it creates a *queue* of vetted projects that are next in line to receive funding when it becomes available.

Benefit #8: Provide Concrete Feedback on Strategy Attainment

A fairly surprising gap in most organizations is that there is no mechanism to measure strategy attainment. There is a lot of effort spent on strategic planning, but a puzzling lack of feedback on how that journey is actually progressing. Sometimes there is a descriptive assessment, often filled with jargon, that does not really help provide a sense of *how much is complete* and *how much is yet to go* in terms of the strategy road map. This is a huge handicap to the CIO, because unless the difference between *actual progress in the strategic road map* and *expected progress in the strategic road map* is known, how can corrective actions be taken in a timely manner?

Of all the benefits provided by the portfolio office to the CIO, this may be the most valuable and strategic in nature. The portfolio office constructs and maintains the strategy attainment map as a derivative of the strategic road map with the overlay of mEVM applied on it (refer to Chapter 11 for a detailed treatment). Using this construct, and overlaying the mEVM data as seen through monthly readouts, the portfolio office is able to create a report that shows the actual strategic attainment as compared to the strategic road map. What makes this even more useful is a comparison of the actual dollars spent while obtaining the strategic attainment. This could be a thought-provoking and far-reaching report for the CIO to see; one that actually informs whether all the cost and effort spent by the information technology organization has actually moved the needle or not.

HOW THE CIO IS THE BIGGEST ENABLER OF THE PORTFOLIO OFFICE

In the previous section, we saw how the portfolio office provides the CIO with a host of benefits, both tactical and strategic. At the same time, the portfolio office is heavily dependent on the CIO for support. The portfolio office needs a powerful patron from whom it derives its power—and that patron is the CIO. Here are the prominent ways in which the CIO can enable the portfolio office to reach its potential and deliver all of the discussed benefits back to the CIO.

Enabling Action #1: CIO Needs to Equip the Portfolio Office with a Mandate

The portfolio office needs to be given a clear, public mandate to effect all of the changes it needs to make in the organization. As noted in Chapter 16, this mandate is key to the portfolio office claiming the role of a change agent. The CIO needs to provide this mandate in a clear, public way and periodically reinforce the mandate. The CIO can also clearly set the tone with public celebrations of the portfolio's successes.

Enabling Action #2: CIO Needs to Fund the Portfolio Office with Resources

The portfolio office needs resources to achieve its full potential. It may include all of the following:

- Hiring a seasoned portfolio manager
- Funding for staffing a small team, potentially temporary staff
- Funding for portfolio consulting services to jump start the effort
- Software licenses for portfolio software (the emphasis in this book has been on low cost SharePoint)
- Training the organization on new portfolio processes
- Change management, including *how-to* content for a website

Enabling Action #3: CIO Needs to Provide Political Cover to the Portfolio

As covered extensively in Chapter 16 concerning politics relating to portfolio management, it is a certainty that the portfolio effort will face opposition from some stakeholders who would prefer the status quo. There may also be missteps on the part of the portfolio team as they attempt to effect the necessary changes. In all of these situations, the CIO needs to come out strongly in support of the portfolio team. The CIO needs to signal to the organization that the portfolio office has CIO support and that the portfolio agenda is not one to be sidelined. There also needs to be an implicit (or explicit) message that stakeholders trying to thwart the portfolio implementation will be held accountable. This may involve one or more high profile terminations of stubborn players whose refusal to comply with portfolio guidelines may threaten to derail the entire effort.

Enabling Action #4: CIO Needs to Be Accessible to the Portfolio Office

Consider a situation where the portfolio office has a mandate, and all of the necessary resources. Can the CIO leave the portfolio office alone and expect all the strategic transformation to happen? No, of course not.

While an empowered portfolio office can do a lot of good by itself, the transformative power of the portfolio lies in this closed loop: Gather Data → Analyze Data → Detect Deficiency→ Advise CIO → Get CIO Decision → Perform Correction → Gather Data.

The CIO (or their proxy, governance) needs to be available to receive the advice from the portfolio office and agree to the recommendation so that the portfolio office can *perform correction*. The CIO needs to act on the advice provided by the portfolio office for the loop to work. To ensure that this happens, there needs to be an arrangement for the portfolio office to have a regular touchpoint with the CIO.

Enabling Action #5: CIO Needs to Appoint Governance Body that Works

The current portfolio trend is for a body of senior, empowered people, collectively called *governance*, to function as the proxy for the CIO and provide oversight to the portfolio. This also optimizes the CIO's time, since governance can filter out the noise and make decisions on behalf of the CIO. However, as frequently seen in many organizations, governance can fall short of the mark and become more like an ineffective *country club* body. When that happens, governance members shy away from making the tough decisions. Other pitfalls include a disinterested governance body or one that doesn't feel adequately empowered. The CIO needs to address these problems in order to prevent the portfolio from falling into dysfunction. First, the CIO needs to appoint governance members that are motivated, knowledgeable, and willing to shoulder the responsibility of providing oversight to the portfolio. Second, the CIO should try to get portfolio training for the governance members—this is overlooked all too often and the portfolio suffers as a result. Third, the CIO needs to review effectiveness of governance periodically and be willing to cycle out members for whom the role is not a good fit.

LEVELS OF PORTFOLIO CAPABILITY MATURITY

Level 1

- The CIO and the portfolio office are isolated from each other, other than the issuance of periodic reports
- The strategic transformation road map is poorly defined, if it exists at all
- If a strategic transformation road map exists, the portfolio office has little or no input on the artifact
- The portfolio office is not empowered by the CIO to choose the right projects and block the wrong ones
- The portfolio office has no say in recommending termination of underperforming legacy projects that continue to occupy organizational resources
- The portfolio office has no connection to the CIO that enables early communication of projects that are going awry

- The portfolio office is not empowered to create a meritocracy that can supersede politics
- The portfolio office is not empowered to carry out annual planning
- The portfolio office is not allowed to play a role in measuring strategic attainment in objective terms

Level 2

- The CIO and the portfolio office have some contact, but it is sporadic and nonuniform
- A strategic transformation road map exists, even if it is not complete and detailed
- The portfolio office does not own the strategic road map—it may have a limited role to play in the review and consumption of the road map
- The portfolio office has input about project intake but still may not be fully empowered by the CIO to choose the right projects and block the wrong ones
- The portfolio office may recommend the termination of underperforming legacy projects, but these recommendations may not be acted upon at all times
- The portfolio office is able to communicate to the CIO about projects going awry; however, the CIO office may not always take the necessary action
- The portfolio office is not able to create a meritocracy that can supersede politics
- The portfolio office is able to carry out annual planning, but it may not be optimized
- The portfolio office is able to play a limited role in measuring strategic attainment in objective terms

Level 3

- The CIO and the portfolio office are in regular contact and aware of each other's perspective
- A complete, detailed strategic transformation road map exists; it is well socialized and maintained/updated regularly
- The portfolio office owns and maintains the strategic road map—it also plays a major role in the organizational review and consumption of the road map
- The portfolio office has complete control over project intake and has been empowered by the CIO to choose the right projects and block the wrong ones

- The portfolio office has the CIO's backing to recommend and follow through on the termination of underperforming legacy projects
- The portfolio office has a hotline to the CIO office to communicate about projects going awry; the CIO office is responsive to such notifications and provides the portfolio office with the necessary support to remediate the situation
- Through mEVM implementation and effective governance, the portfolio office is able to create a meritocracy that can supersede politics
- The portfolio office is empowered to orchestrate effective annual planning each year
- The portfolio office is able to play a major role in measuring strategic attainment in objective terms

CHAPTER SUMMARY

This chapter explores the critically important relationship between the CIO and the portfolio office. We first explored how the mandate of the CIO creates a direct dependency with the portfolio. We then discussed at length how the CIO is the biggest beneficiary of the portfolio office. We went on to illustrate how the CIO is also the biggest enabler for an optimally functioning portfolio office. We elaborated on this theme by listing all of the services that the CIO can expect from the portfolio office. Next, we listed all of the actions that the CIO can take to empower the portfolio. We conclude the chapter with the indicators found at each level of this portfolio capability.

19

FINANCE

INTRODUCTION

Portfolio management can be thought of as a semi-finance function. The whole endeavor of portfolio management focuses on managing significant sums of the organization's money toward improving the company's capabilities and ensuring that the spending of money stays on track. Therefore, it is quite logical that successful portfolios have a close relationship with Finance as well as support from that function in their organization. In this chapter, we explore the following aspects of the relationship between Finance and the portfolio office:

1. Explore interface points between Finance and the portfolio office
2. The different roles that Finance plays in complementing the portfolio office's function and ensuring strategic success
3. The benefits that Finance derives from the portfolio office

INTERFACE POINTS BETWEEN FINANCE AND THE PORTFOLIO OFFICE

The *touch points* between Finance and the portfolio office are multifaceted and illustrate how tightly these two functions are bound together.

Touchpoint #1: Guidance about OPEX and CAPEX

Capital expenditures (CAPEX) and operating expenditures (OPEX) are two different categories of business expenses.[1] CAPEX are the funds that a business uses to purchase major physical goods or services to expand the company's abilities to generate profits. These purchases can include hardware (such as printers or computers), vehicles to transport goods, or the purchase or construction of a new building. The type of industry a company is involved in largely determines the nature of its CAPEX.

OPEX result from the ongoing costs that a company pays to run its basic business. In contrast to CAPEX, OPEX are fully tax-deductible in the year they are made. As OPEX make up the bulk of a company's regular costs, management examines ways to lower them without causing a critical drop in quality or production output. Sometimes an item that would ordinarily be obtained through CAPEX can have its cost assigned to OPEX if a company chooses to lease the item rather than purchase it. This can be a financially attractive option if the company has limited cash flow and wants to be able to deduct the total item cost for the year.

In a few cases, it is straightforward to distinguish between OPEX and CAPEX; however, in most cases, guidance from Finance is needed to decide whether it is OPEX or CAPEX. Having this guidance and partnership in place is strongly recommended because it could make a major difference in classifying project spend correctly and could potentially save the organization millions in tax management.

Touchpoint #2: Partnership during Annual Planning

As covered in Chapter 2, annual planning is a comprehensive exercise to plan next year's activity. As part of this activity, all of the known demand is gathered and prioritized. This demand is then compared against the available funds and the *affordability line* is drawn where the funds run out. The projects above the line get approved for funding and the projects that are below the line are placed in a queue for future consideration. Finance has two important roles to play during an annual partnership exercise:

1. During the gathering of demand, input from Finance is essential to categorizing that demand as OPEX or CAPEX. This is important in order to get a true picture of demand and avoid surprises later in terms of mismatch between supply and demand. (Consider a situation where the organization realizes later in the year that much of its annual CAPEX demand was actually OPEX, which is now scarce to find.)

2. During the *drawing of the affordability line*, Finance needs to specify how much funding is available for the portfolio next year. This is usually a by-product of the larger financial planning exercise that is done for the whole company. Finance can also advise if any extra funds can be found for strategically important projects which might fall *below* the line.

Touchpoint #3: Benefit Management

As covered in Chapter 7, there needs to be a standard taxonomy on benefits that all stakeholders can agree to follow. Finance's input is crucial when it comes to defining what the different kinds of project benefits are. In the case of hard benefits, Finance needs to sign off that those numbers are valid and are expected to make a difference on the balance sheet. This formal screening by Finance goes a long way in avoiding benefit inflation by project owners.

During benefits management, Finance needs to be on the governance body that reviews benefit performance and compares it to what was promised at the start of the project. As the steward of the organization's money, Finance is expected to take a hard line and recommend terminating projects that are not delivering benefits per plan.

Touchpoint #4: Reporting Actuals on Projects

One of the most vital services performed by Finance is to ensure that the portfolio office has visibility to the actual spend by each project per month. This data is crucial for the portfolio office to make decisions regarding the project. Only Finance can authoritatively produce this *official* information directly from the financial reporting systems. This data is then used for portfolio performance monitoring including modified earned value management (*mEVM*).

Touchpoint #5: Managing Underspend and Overspend

Sometimes a portfolio will, despite best efforts, either overspend or underspend its budget. Sometimes this may even happen for valid reasons—a strategically important project within the portfolio may discover additional scope worth doing that costs more money but is ultimately beneficial to the organization. In these cases, a close working partnership with Finance may save the day—Finance has visibility as to the spending trend of the rest of the organization and may be able to arrange for a way to match the portfolio over/underspend with a corresponding under/overspend in another area of the organization. For this arrangement to work, the portfolio needs to also give Finance sufficient advance notice of an anticipated over/underspend at the end of the year.

Touchpoint #6: An Authoritative Voice on Governance

Finance is an essential part of portfolio governance and must be represented in all key decisions regarding the portfolio. Sometimes Finance needs to exert its role as the steward of the organization's funds and ensure that the right decisions are made in the portfolio. This is key in terms of terminating bad investments and not approving funding for projects that do not have strong benefits or return on investment.

A SYMBIOTIC RELATIONSHIP

As we saw in the previous section, Finance extends many services to the portfolio office that are integral to the smooth working of the portfolio management function in the organization. However, this is a mutually beneficial relationship—a well-run portfolio provides Finance with the assurance of managing project spend in a responsible manner. Predictability in project spend is generally prized by Finance because any adverse surprises in this area (or any other area, for that matter) creates a problem for Finance in managing the overall Financial picture for the enterprise.

Finance can also be a major beneficiary of a well implemented mEVM system. Through mEVM, project performance and spend trends can be objectively captured, enabling Finance to become ready for year-end surprises in over- or underspend.

The foundation for this symbiotic relationship needs to come from executive management, typically in an arrangement worked out between the leaders of information technology and Finance—the chief information officer and the chief financial officer. This understanding is then translated to mid-level management support between Finance managers supporting the portfolio manager.

HOW FINANCE BENEFITS FROM THE PORTFOLIO OFFICE

- **Benefit #1**: Without an effective portfolio office to manage projects, Finance would have to deal with projects running wild in terms of both overspend and underspend. The biggest benefit that comes with an effective portfolio office is the predictability of spend of projects and to a certain extent, an ability to control the spend.
- **Benefit #2**: A portfolio office is able to exert control to ensure that funds are being well spent. As seen in Chapter 10, it is possible to show the multi-year trend in how the organization is getting more efficient at spending money to achieve project objectives. Such an objective indicator of money being well spent would be of interest to Finance leadership in their ongoing role of managing the organization's money.
- **Benefit #3**: The portfolio office monitors the performance of projects and takes a closer look at projects that are not performing well. If an

underperforming project does not recover and continues to show a deteriorating trend, the portfolio office will move to terminate the project. Such prompt action is necessary to ensure that bad projects do not crowd out the good projects and waste the organizational resources. Finance relies on the portfolio office to make these hard decisions and ensure better use of the organization's funds.

LEVELS OF CAPABILITY MATURITY

Level 1

- There is no concept of partnership between the functions of the portfolio office and Finance.
- The organization may not have an effective framework to treat OPEX and CAPEX differently.
- Even if the organization classifies OPEX and CAPEX separately, Finance may not offer sufficient support to the portfolio office in distinguishing the two classes of funding.
- There may be little partnership between the portfolio and Finance during annual planning. This translates into a situation where project demand and funds supply have no relation to each other, and results in diminished confidence in the utility of the annual planning exercise.
- There may not be any coordination between the two functions on benefit management. This results in projects making promises of generous benefits without any basis to back them up.
- In the absence of an effective partnership with Finance, the portfolio office may not have a standard, systemic method of obtaining project actuals. This greatly hampers the function of the portfolio office.
- Due to the lack of communication and partnership between the portfolio office and Finance, there may not be a view to anticipate and control overspend and underspend at the end of the year.
- Finance may not be represented on the portfolio governance body.
- mEVM may not be implemented in the organization. Therefore, there is no way for the portfolio office to create objective performance data indicators to share with Finance.
- Due to lack of alignment between the two functions, there may not be mid-level management support (such as a Finance manager dedicated to supporting the portfolio office).

Level 2

- There is awareness of the need for partnership between the functions of the portfolio office and Finance. Accordingly, the two functions endeavor to coordinate and align wherever possible.
- The organization has an effective framework to treat OPEX and CAPEX differently at all levels, including project spend. Finance offers authoritative guidance and support to the portfolio office (and projects) in distinguishing the two classes of funding.
- There is an effective partnership between the portfolio office and Finance during annual planning. Therefore, project demand and funds supply are managed to be within range of each other, and this arrangement creates confidence in the utility of the annual planning exercise.
- There is coordination between the two functions on benefit management. This results in projects being held accountable for promised benefits.
- Due to the partnership with Finance, the portfolio office is able to depend on a standard, systemic arrangement of obtaining project actuals. This significantly helps the function of the portfolio office.
- Due to the coordination between the portfolio office and Finance, there is some visibility regarding the anticipated overspend and underspend at the end of the year.
- Finance is represented on the portfolio governance body.
- mEVM may not be implemented in the organization. Therefore, while there is sufficient partnership between the two functions, there is still the lack of objective performance data indicators—such as mEVM—that would form the basis of decisions.
- Due to alignment between the two functions, day-to-day support is available at mid-level management (such as a Finance manager dedicated to supporting the portfolio office).

Level 3

- There is awareness of the need for partnership between the functions of the portfolio office and Finance. Accordingly, the coordination and alignment between the two functions are optimized to a high degree.
- The organization has an effective framework to treat OPEX and CAPEX differently at all levels, including project spend. Finance offers authoritative guidance and support to the portfolio office (and projects) in distinguishing the two classes of funding.
- There is an optimized partnership between the portfolio office and Finance during annual planning. Therefore, project demand and funds

supply are matched to a close degree, and this arrangement creates confidence in the utility of the annual planning exercise.

- There is coordination between the two functions on benefit management, with Finance taking a leadership role on the Benefits Review Council. This results in each project's benefit statement being screened for viability before the project is approved and the project then being held accountable for promised benefits.
- Due to the partnership with Finance, the portfolio office is able to depend on a standard, systemic arrangement of obtaining project actuals. This significantly helps the function of the portfolio office.
- Due to the coordination between the portfolio office and Finance, there is heightened visibility to anticipated overspend and underspend at the end of the year. This visibility is enhanced by the presence of a robust mEVM implementation.
- Finance is represented on the portfolio governance body and weighs in on all key decisions.
- mEVM is comprehensively implemented in the organization. This provides an objective basis for the partnership between Finance and the portfolio office.
- Due to alignment between the two functions, day-to-day support is available at mid-level management (such as a finance manager dedicated to supporting the portfolio office).

CHAPTER SUMMARY

This chapter explored the vital partnership between Finance and the portfolio office. We first explored the multiple interface points between Finance and the portfolio office and illustrated how tightly the functions are woven together. We then listed all of the different benefits that the portfolio office receives from partnering with Finance. To show the symbiotic relationship between the two functions, we went on to explore all of the benefits that the Finance office receives from working with the portfolio office. We concluded the chapter with the indicators found at each level of this portfolio capability.

NOTE

1. *Investopedia, The Difference between CAPEX and OPEX,* (http://www .investopedia.com/ask/answers/020915/what-difference-between-capex -and-opex.asp)

20

PORTFOLIO ROLLOUT AND
CHANGE MANAGEMENT

INTRODUCTION

Many organizations launch portfolio management with lots of fanfare but
there is little adoption or observed throughput in the months that follow. What
could be the problem? In many cases, it could be a lack of change management,
including adequate training, during rollout. Change management can guide the
organization in effectively navigating the change created by the introduction of
the new system of managing projects and programs. In this chapter, we'll review
the following aspects of portfolio rollout and change management:

1. Review the range of situations that provide the context for a portfolio
   rollout
2. Analyze why portfolio rollout efforts often fail
3. Discuss the politics surrounding portfolio rollout and change management
4. Discuss recommended best practices for portfolio rollout
5. Introduction to the mind map
6. Typical mind map structure for a portfolio office
7. Success factors for a portfolio mind map
8. Levels of portfolio maturity for this capability

THE WIDE SPECTRUM OF PORTFOLIO ROLLOUTS

It should be kept in mind that the term *portfolio rollout* can span a wide range
of situations. In some organizations, there may be no portfolio management at

all—therefore this rollout may constitute the first contact with a totally new concept. At other organizations, there may be some kind of portfolio management in place, and the present rollout may be an improvement or enhancement—possibly an attempt to go from Level 1 to Level 2 or even from Level 2 to Level 3. At other places, there may have been some kind of unsuccessful attempt in the past to implement portfolio management and this rollout is a *do-over*. Each of these situations comes with their own challenges and will have to be handled accordingly through effective change management. Although change management is an established discipline with specialized (and expensive) consultants, sometimes a back-to-the-basics approach can deliver the desired results.

WHY DO PORTFOLIO ROLLOUT EFFORTS OFTEN FAIL?

The following information includes some of the reasons why portfolio rollouts have faltered.

Reason #1: Inadequate Estimation of Impact

In the minds of portfolio managers, the changes they roll out seem simple, welcomed, and straightforward. They may also seem to be a big improvement from the current suboptimal state of affairs. What they may not realize is that the organization may have adapted, however awkwardly, to the confusion or suboptimal situation. Any change now, even if for a better outcome, may create discomfort and uncertainty and is not likely to be warmly received. A prudent portfolio manager would do well to factor this potentially chilly reception while designing the rollout.

Reason #2: Incomplete Analysis of Affected Parties

It is important to methodically map out all of the current stakeholders and plan for how they would be affected by portfolio rollout. It is also important to consider the new stakeholders who would now come under the purview of portfolio management as a result of this rollout. Successful rollouts create a formal matrix of stakeholders, describe the expected impact for each stakeholder group, and also have a checkmark to show that each stakeholder group was helped in dealing with the rollout. This matrix is also useful to show the organization that the needs of all stakeholders were factored for the rollout and taken care of.

Reason #3: Inadequate Customization

Since the portfolio rollout affects different stakeholders to different degrees, it would be a mistake to provide the same level of training to everyone. Consider a project manager/coordinator who needs to be fully trained on the new templates, where to find them, where to submit them, etc. At the same time, a project team member at least needs minimal training—perhaps just an awareness that all projects are now managed as part of a portfolio. Providing the full extent of training to all is a waste of time and may even leave everyone feeling that *the new portfolio procedures are too complex*. Therefore, customization is required to provide the appropriate amount of training for the varied groups of stakeholders. Customized training, in turn, relies upon adequate impact estimation and stakeholder analysis as covered in Reasons #1 and #2.

Reason #4: Lack of Scenario-Based Training

Many portfolio rollout trainings miss the mark because they launch into a detailed explanation of the new portfolio policies and procedures that, while correct, are not really relevant to the day-to-day functioning of the stakeholders. It's much more useful for the training to focus on how the actions performed by the stakeholders today would change in the new process. For example, how does a stakeholder submit a project proposal for funding consideration? How does a project request more money? Etc. The section on mind maps later in this chapter addresses this need for delivering scenario-based help to the organization in detail.

Reason #5: Lack of Distinction from Earlier Failed Rollouts

In some organizations, prior rollouts of portfolio changes may not have been successful and may have created unfavorable opinions in the collective organization. In such situations, where an earlier rollout of portfolio management was attempted with mixed results, it becomes important to emphasize why this rollout is different. It may also help to frame this enhancement effort as a response to the shortcomings of the previous attempt to roll out portfolio management.

Reason #6: Attempting Too Much Change

Every organization has a finite *change management capacity*—an ability to understand and process change. By trying to introduce too much change at once, the portfolio office runs the risk of failure. What are some examples of trying to do too much? For instance, trying to introduce a new enterprise portfolio management software at the same time as rolling out significant changes to the portfolio process that is currently in place. This would likely strain the organization and cause pushback and poor adoption of the whole setup. The recommendation would be to roll out the portfolio process first, let it mature for a bit, and then try to roll out the portfolio tool.

Reason #7: Failure to Account for Users' Lack of Retention

Sometimes an organization will launch an effective change management effort around a new portfolio process rollout. The change management and training could be well attended and well received; however, inexplicably, the stakeholders may indicate dissatisfaction with the whole process a few months later. What could have gone wrong? The answer may lie in the availability of post-rollout resources.

It's a well-known fact that people have limited retention of training content that they may not apply in practice immediately. For example, people may forget how to initiate a new project proposal unless they have done it a few times—and they may not get a chance to do it for some time after their training. It's very important to complement an effective training course with resources such as a training website that people can refer to. It is also essential to make this training website an easy to use resource, as seen in the following section.

THE POLITICS OF PORTFOLIO CHANGE MANAGEMENT

As we covered in Chapter 16, everything about portfolio management can be political. This is due to the fact that a high-functioning portfolio forces accountability on all players. Stakeholders whose subpar performance was previously masked by the chaos now have to contend with high visibility and oversight from the portfolio office.

Stakeholders who are not supportive are often looking for the portfolio office to slip up and give them a chance to criticize the whole portfolio setup. A botched rollout would serve as the perfect opportunity for such a maneuver—a stakeholder could claim that the training was inadequate and hence call into question the entire competence of the portfolio office. By extension, they would also claim some lenience or a degree of exemption from following the portfolio procedures. How can this be dealt with? Some options are outlined here:

1. The recommended approach is to create a comprehensive stakeholder analysis to ensure that all stakeholders are accounted for. The best way

to do this tactically is to prepare a matrix of stakeholders, accompanied by their official role, and list the impact to them caused by the roll out of portfolio management.

2. It helps to circulate this matrix to the whole organization (or as broadly as needed), asking to validate this matrix, and identify any additional stakeholders that may have been missed.

3. Further, during the actual training sessions, it's useful to capture which members from which team attended and then circulate this attendance record as part of the change management communications.

Collectively, the approach just described prevents sniping from certain teams that may try to claim that they *weren't trained*. The portfolio office can simply refer to the roster—one can either prove that many of the complaining team's members attended training or that many members did not attend, despite an invitation. In either case, this proves that the portfolio office is blameless.

RECOMMENDED BEST PRACTICES FOR PORTFOLIO ROLLOUT

Best Practice #1: Delivering Effective Training

Effective training is the product of careful design and well-crafted delivery. The portfolio office should make every effort to ensure that the training is relevant, concise, and designed to fit the users' needs. It would also help if the training itself were delivered by professional trainers. Too often, portfolio offices tend to rely on the in-team talent to deliver the training content, which may deliver underwhelming results in terms of audience engagement and retention. It is important that the portfolio office recognize the importance of the training/ roll-out exercise and engage professionals to deliver the service.

Best Practice #2: A Network of Local Champions

Only a small percentage of people who attend a training session grasp everything that is imparted, record all relevant information, and remember to apply it correctly when the situation arises. This is all the more true for areas where the received training is not applied immediately. When the need does arise, most people would prefer consulting a person and being walked through the entire procedure, at least for the first time. A local champion is a person who fulfills exactly that need.

In this case, it would be a person—probably a project manager or project coordinator—who received intensive training and is able to function as an expert within their designated department. This person would be able to walk a project owner or any other stakeholder through any relevant portfolio procedure and give them the reassurance of working with a cohesive, functional system.

Best Practice #3: Website with Archived Training Materials

Imagine a successful portfolio rollout accompanied by effective training. Workshops are held and training materials are produced and distributed. What's the next logical step to make this a comprehensive effort? Archiving the training materials, including recordings of the training, and posting them on an easy-to-navigate website, enabling users to review them at a later date when they actually need to put the training concepts into practice. This arrangement also benefits new employees in the organization who joined after the comprehensive trainings were held.

Best Practice #4: Intuitive Navigation of the Portfolio Process and Materials

One of the most common complaints heard among the rank and file of an organization is, "I can't find what I need to get the job done." This is particularly true of portfolio management, where there are a lot of process materials, templates, and historical data that tends to pile up, creating a maze that is hard to navigate. This one factor could make all the difference between the organization adopting portfolio management processes or avoiding it and trying to work around it.

How do we balance these competing demands? On the one hand, we need to preserve and store all relevant data on the portfolio website, creating a huge pile to sort through. On the other hand, we need to get the right information to the right person as quickly as possible, enabling them to transact their portfolio work. One approach to accomplish this task is through the use of a mind map.

WHAT IS A MIND MAP?

A mind map is a diagram used to visually organize information. A mind map is hierarchical and shows relationships among pieces of the whole. It is often created around a single concept, drawn as an image in the center of a blank page, to which associated representations of the central concept are added. Major ideas are connected directly to the central concept, and other ideas branch out from those. The utility of a mind map in representing information lies in its similarity to the human thought process, which involves starting with a broad outline and branching to the highly specific. The next section explains in detail how to use a mind map to ensure that the organization is able to navigate, locate, and use portfolio materials as part of the portfolio process.

HOW TO USE THE MIND MAP

Figure 20.1 shows a basic example of a mind map that helps partners of the portfolio office navigate through the mass of information. The big oval in the center stands for the *core idea* or *central topic* of portfolio management. Surrounding the central theme of *portfolio management* are the major topics that a typical stakeholder would be concerned with, including:

1. **New project initiation**: A stakeholder who wants to start a new project would start here and be guided on how to start a new project in accordance with portfolio policy. There would be resources on what the process is to start a new project, as well as how to use templates and other resources that would be relevant in enabling the stakeholder to submit a new project.

2. **Existing project maintenance**: A stakeholder who already has a current project in the portfolio would find this link to the logical starting point to transact with the portfolio. There would be resources on reporting the performance of the project back to the portfolio and staying in compliance with the other portfolio processes for an existing project.

3. **Historical portfolio data reference**: A person wanting to refer to historical portfolio data would find this link to be the ideal place to start looking for the particular data that they are interested in. Historical data could cover portfolio decisions, including funding and other approvals. It could also contain historical performance data.

4. **Modified earned value management (*mEVM*) portal**: This link would be the gateway for all matters related to mEVM. This would include mEVM artifacts for each project, such as the current year's month-by-month snapshots of mEVM performance. All other mEVM reports such as aggregation would also be found here.

5. **Annual planning portal**: This link would be the starting place for all topics related to the annual planning exercise. The process, the templates, and all of the artifacts related to annual planning would be accessed from this point.

6. **Training materials archive**: This link would be the gateway to all training materials, including archives of previously conducted training sessions. This is a valuable resource for people who could not attend training sessions or who need refreshers on the training content.

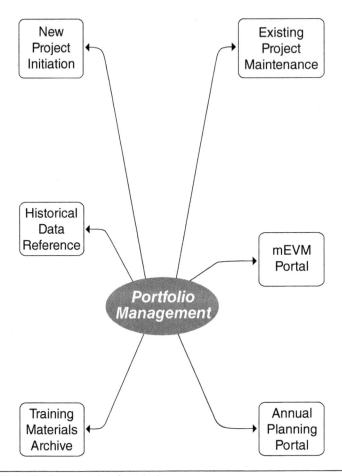

Figure 20.1 Basic example of a mind map

The aforementioned list represents a very simple, first-order of topics surrounding the central topic. However, for these topics to be useful, they need to be elaborated upon. For example, a user wanting to start a new project would expect further links specifying the correct type of project and the corresponding template that the user would need to fill out. Accordingly, the previously listed six topics are elaborated on in the following sections, with an exploration of what subtopics would fit under each topic:

Elaboration of Topic #1: New Project Initiation

This topic would be the landing point for all users who need to access resources involved in starting a new project for inclusion in the portfolio. Here are the correlating subtopics:

a) **Subtopic #1—New project template**: For a user looking to start a new project, perhaps the most important resource would be the new project template. This topic would contain links to the new project template as well as instructions to fill out the template. It would also contain instructions on how and where to submit the template.

b) **Subtopic #2—Intake process guide**: The next subtopic of interest to the user under this main topic would be an explanation of the process that governs approval of new projects. This may involve a process flow, explanation of the roles of the different parties involved, and the entry criteria for a new project. For example, new projects that are larger than a particular threshold (say $100K) would go through a different process than one that is below the threshold.

c) **Subtopic #3—Governance timetable for approving a project**: Can a user submit a new project at any time? In all probability, they cannot do that and the timetable for considering/approving a new project proposal is driven by the overall portfolio governance cadence. Therefore, a typical user would find it valuable to understand the governance calendar and would find it useful to see this subtopic under this main topic.

The elaborated branch of the mind map for Topic 1 is shown in Figure 20.2. It needs to be kept in mind that the subtopics are usually hyperlinks that, when clicked upon, take the user to the corresponding web page with the relevant information and content. Also, it's good to remember that the mind map is not duplicating the information—it is only providing a logical and intuitive navigation path to the content that is already there.

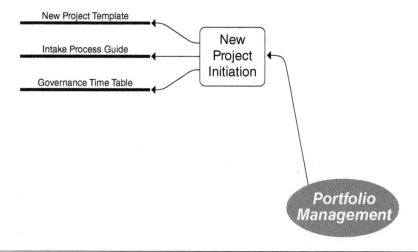

Figure 20.2 Elaborated branch of a mind map for new project initiation topic

Elaboration of Topic #2: Existing Project Maintenance

This topic would be the landing point for all users with an already approved and currently running project. Such users would still need to come back to the portfolio for various actions such as reporting on the performance of their project, to request additional funds, or to return excess funds. They may also need the portfolio's permission to allow changes to their officially declared scope or schedule. Here are the subtopics that would offer further help for such users:

a) **Subtopic #1—Existing project status template**: The most common situation for an existing project is to come to the portfolio periodically to deliver a status update. This subtopic handles that situation with links to the existing project status template. It would also be helpful for this subtopic to be accompanied by another subtopic with instructions on how to fill out the template.

b) **Subtopic #2—Additional funds request template**: The next most common situation for a project is to come to portfolio governance to ask for more money. This subtopic handles that situation with links to the additional funds request template. Ideally, there would be another subtopic with instructions on how to fill out the template.

c) **Subtopic #3—Scope/schedule/budget change request template**: Projects are sometimes faced with the need to change their vital parameters such as scope, schedule, and budget. In a well-governed portfolio, these changes need to be approved by portfolio governance. Therefore, an existing project coming to portfolio governance to request a change would need to use the standard template to present the request. This subtopic handles that situation with links to the scope/schedule/budget change request template. (Two options are possible here: there could either be one common template for scope, schedule, and budget—or there could be separate templates for each. If the templates are separate, there would be three different subtopics, with each containing a link to the corresponding template.) There would also be another subtopic with instructions on how to fill out the template or templates.

d) **Subtopic #4—Funds turnback template**: In a well-run portfolio, projects are encouraged to return money that they are unlikely to use. This money can then be allocated to other projects in order to generate a more significant strategic impact. This subtopic handles that situation with links to the funds turnback template. It would also be helpful for this subtopic to be accompanied by another subtopic with instructions on how to fill out the template.

e) **Subtopic #5—Governance timetable for existing project requests**: Can an existing project approach governance at any time with their requests or transactions? It's unlikely because governance bandwidth is limited and there are many projects that are in need of a review and decisions by portfolio governance. To accommodate everyone, there would be a timetable for portfolio governance that would be in line with the overall portfolio governance cadence. Therefore, a typical user who is planning to come to the portfolio with requests would find it valuable to understand the governance calendar and would find it useful to see this subtopic under this main topic.

f) **Subtopic #6—mEVM update**: mEVM is a powerful technique that serves as a catalyst for improving portfolio performance. Although the technique is simple, it calls for monthly updates showing the progress made on the project, in addition to updates made to project financials, specifically project spend. Arrangements need to be made such that a project manager running an existing project can quickly get to their mEVM template and update it each month. Enabling easy access to mEVM artifacts is critical to ensuring that all projects provide their mEVM updates each month. Accordingly, this subtopic would contain links that let the user navigate to the mEVM artifacts that are specific to their project.

The elaborated branch of the mind map for Topic 2 is shown in Figure 20.3.

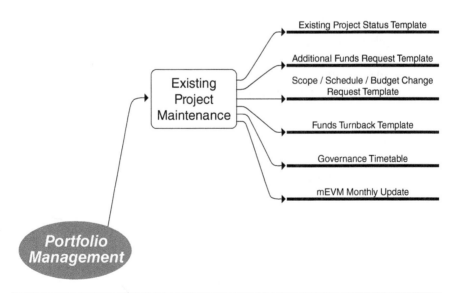

Figure 20.3   Elaborated branch of a mind map for existing project maintenance topic

Elaboration of Topic #3: Historical Portfolio Data Reference

A portfolio generates a significant amount of data over time. While it can be quite daunting to sort through this maze of data, it must be kept in mind that this historical data could offer valuable insights to a project manager running current projects. Accordingly, the portfolio office needs to enable the project managers and any other stakeholders who need to refer to this data. The following subtopics would help the user to navigate further and retrieve the specific data that they are interested in:

a) **Subtopic #1—Historical project artifacts**: Artifacts related to past projects can be very useful as a starting point for drafting new artifacts for present projects, or just as a reference. For example, old project plans may offer a lot of utility in jump-starting the creation of a new project plan. Therefore, it's a good idea to keep these organized and available for reference. This subtopic provides links to navigate through past projects and arrive at the exact artifact that the user is interested in.

b) **Subtopic #2—Historical portfolio artifacts**: Just like previous project artifacts, previous portfolio artifacts are equally useful as a starting point for a project manager trying to draft materials for a portfolio appearance. For example, it may be useful to refer to a previous project proposal while creating a new one. Or it may be expedient to take an old status update and make minor changes to arrive at the current status update. To enable all of the above transactions, this subtopic provides links to navigate through all of the previous periods' portfolio artifacts and obtain the one the user is looking for.

c) **Subtopic #3—Historical lessons learned**: Projects that were closed out (successfully or otherwise) have a lot of insight for project owners who are trying to start new projects. Some projects actually perform a formal *lessons learned* exercise, which provides even more information on things to do and things to avoid. Therefore it is very important to preserve these in an organized fashion and also enable these to be searchable and accessible to stakeholders who are looking for this information. To enable all of the above accessibility, this subtopic provides links to navigate through all of the previous projects' closeouts and lessons learned.

d) **Subtopic #4—Historical governance decisions**: Some long-running projects may have a life that stretches across several years. Such a multiyear history may have several key governance decisions that need to be revisited and referenced at some point. The key system for this purpose was explained in detail in Chapter 15, namely the historical portfolio actions repository (HPAR). The links to HPAR would be found under this subtopic.

The elaborated branch of the mind map for Topic 3 is shown in Figure 20.4.

Elaboration of Topic #4: mEVM Portal

The mEVM technique creates a huge amount of data each month, which adds up over the years. This data is needed for reference as well as for trend analysis. This main topic would be of interest to any user whose project follows mEVM. All mEVM-related information would be found at this link and the user would click on the following subtopics to transact further:

a) **Subtopic #1—mEVM process and training materials**: Most people are new to mEVM and would find it reassuring if they could access a ready reference to the process. They would also find it useful to review the training materials a few more times after receiving the original training. To provide the users with this facility, this subtopic includes links to the mEVM process and the training materials.

b) **Subtopic #2—mEVM templates**: mEVM templates are integral to the mEVM technique and are sought after by the users who are practicing mEVM. The mEVM templates could also be tweaked and improved over time—and giving the users a way to always access the latest official version is a good idea. This subtopic would contain links that would point the users to the latest templates.

c) **Subtopic #3—Monthly mEVM artifacts**: The whole mEVM system is based on being easy to use and not taking up too much time to update. Accordingly, users find it easy to use last month's mEVM artifacts to

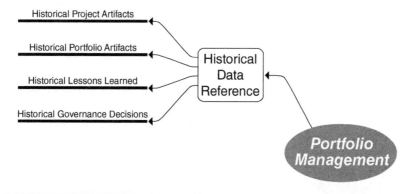

Figure 20.4   Elaborated branch of a mind map for historical portfolio data reference

make minor changes that represent the delta from the previous month to the current month. This modified artifact is then submitted as the mEVM update for the current month. To enable the speedy update process just described, project managers need to be able to access the previous months' mEVM artifacts belonging to their specific project. This subtopic would contain links that would point the users to their projects' mEVM submissions from previous months.

d) **Subtopic #4—Aggregated mEVM data**: mEVM data is stackable, which allows it to form insightful composite views of the portfolio's information. This ability to aggregate allows the portfolio office to provide views that switch between the big picture and the details, giving portfolio governance visibility across all dimensions of the portfolio. Aggregation of mEVM also allows visualization, analysis, and trend prediction at various levels of granularity across the portfolio. This subtopic would contain links that would point the users to the different views created by mEVM aggregation.

e) **Subtopic #5—Strategic attainment using mEVM**: mEVM makes it possible to measure strategic attainment, which has always been a sought-after goal for portfolio offices as well as executive leadership. By making a few design tweaks to the mEVM artifacts and applying them to the task of measuring strategic milestones, reports can be generated that show strategic attainment. These reports can be accessed concerning this topic through links contained in the subtopic.

The elaborated branch of the mind map for Topic 4 is shown in Figure 20.5.

Elaboration of Topic #5: Annual Planning Portal

Annual planning is another portfolio activity that generates a large quantity of data that will need to be referenced multiple times. For annual planning to be

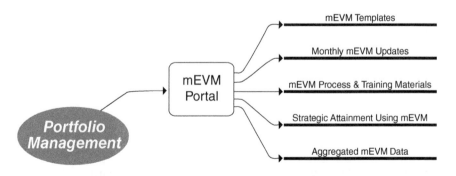

Figure 20.5   Elaborated branch of a mind map for mEVM portal

successful, all of the users in the organization need to have easy access to the annual planning materials. This link would contain all of the annual planning related materials and data. A user who needs to transact any actions related to annual planning would start at this main topic and find the following subtopics:

a) **Subtopic #1—Annual planning artifacts**: Annual planning can be a complex and long-drawn-out exercise with several important artifacts, the most important of which is the annual planning template. There are also other artifacts that the user would need to refer to during the process, such as the bubble chart and other compilations of the collected annual demand. It would be a huge help for all participants of the annual planning effort to know that they can access all of the artifacts related to the annual planning effort in one place that is easy to navigate. The facility of easy access to the artifacts is provided through links contained in this subtopic.

b) **Subtopic #2—Annual planning training materials**: Although annual planning is launched with a comprehensive training session, the reality is that most people can recall little or nothing about the training when they finally start working on their annual planning demand templates. Another factor to keep in mind is that annual planning can be a months-long effort, and even people who retained much of the training in the beginning may want a reference to refresh their memories. Given all of these factors, most users in an organization would find it useful to have a single place from where the annual planning training materials can be accessed. This subtopic serves that exact need by containing links that point the user toward the annual planning training materials.

c) **Subtopic #3—Historical annual planning submissions**: Annual planning continuity needs to be preserved and the effort spent in each annual iteration needs to be optimized. The goal is to ensure continuity from one year's annual planning to the next, while keeping each year's effort as low as possible. The key to running annual planning with minimal effort is to preserve the previous year's annual planning submissions. Starting from last year's submissions and only managing the changes from year to year will allow for a more modest and easier effort each year. This subtopic serves that critical need by containing links that point the user toward the previous year's annual planning submissions.

d) **Subtopic #4—Annual planning strategy road map**: For annual planning to be effective, the exercise needs to be underpinned by a strategy road map. However, the problem in most organizations is that the rank and file who initiate projects have no knowledge of the strategic road map. To remediate this issue, the organization needs to place a well-socialized strategy road map within easy reach of project owners who are involved in the annual planning. This would vastly simplify the annual planning process and succeed in matching proposals to key drivers of strategy. This subtopic serves that critical need by containing links that point the user toward the accepted strategy road map of the organization.

e) **Subtopic #5—Historical annual planning funding decisions and queue**: Annual planning always occurs in a continuum. Projects that were considered for funding but then fell *below the line* in one year's annual planning exercise are added to the queue and then considered for funding throughout the next year—and if still not funded, they are considered for the next year's annual planning. Accordingly, it may be useful for a department to refer to last year's funding decisions and use that input to create the list of projects to bring forward as this year's demand. This subtopic serves that need by containing links that point the user toward the historical annual planning funding decisions and (last year's) queue.

The elaborated branch of the mind map for Topic 5 is shown in Figure 20.6.

Elaboration of Topic #6: Training Materials Archive

Why should there be a main topic of training materials? After all, almost every other topic has a subtopic dealing with training and the archival of the training

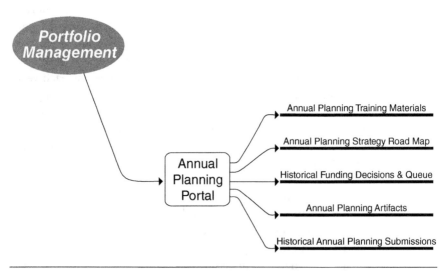

Figure 20.6 Elaborated branch of a mind map for annual planning portal

materials related to that topic. The reason why there is a need for a main topic related to training is simply to allow the user to access the training resources in a straightforward manner. Rather than click on several main topics and navigate to the subtopics of each topic while trying to locate training, it would be far more intuitive to go to a main topic called *training* and then navigate once to the kind of training the user is interested in. The following subtopics would be appropriate under this main topic:

a) **Subtopic #1—Basic training in the portfolio management process:** The majority of people in an organization have only basic transactions to conduct with the portfolio office. Essentially, all that these people need to know is that there is a portfolio that oversees the initiation, execution, and closeout of projects. For those people, the essentials of interacting with the portfolio office are captured in the *basic training* series. This subtopic serves that need by containing links that point the user toward the basic training materials about the portfolio management process.

b) **Subtopic #2—Advanced training in the portfolio management process:** Although the majority of the people in an organization have only a basic relationship with the portfolio office, there are some people— including project managers and project owners—who have extensive dealings with the portfolio office. These people need to know in detail how to submit a project, how to participate in annual planning, how to provide periodic updates to the portfolio office, and so on. The training that imparts the knowledge of all of these detailed transactions are captured in the *advanced training* series. This subtopic serves that need by containing links that point the user toward the advanced training materials about the portfolio management process.

c) **Subtopic #3—mEVM training:** mEVM is a new concept for most people in an organization and hence, needs extensive training during the initial rollout. However, the need for training remains after the initial rollout—new employees in the organization need to be trained in mEVM and existing employees would also need a refresher in mEVM concepts from time to time. For all of those reasons, the mEVM training needs to be archived and available at one accessible location. This subtopic serves as the location that contains links that further point the user toward the specific mEVM training materials.

d) **Subtopic #4—Annual planning training:** Annual planning is a huge and complex exercise that can last for several months. Although each yearly session of annual planning kicks off with a training session, most people have to wait for a while before implementing the concepts taught in the training session. It's quite understandable then that people would want a ready reference to the annual planning training material so they can refresh their understanding. This subtopic serves as that ready reference point that contains links that further point the user toward the annual planning training materials.

The elaborated branch of the mind map for Topic 5 is shown in Figure 20.7.

This concludes a walkthrough of a typical mind map that allows the user base of an organization to navigate the complex portfolio management content of the enterprise's portfolio office. The complete mind map for the whole portfolio is shown in Figure 20.8. Keep in mind that some subtopics can belong in more than one topic. That is perfectly alright because the ultimate aim is to let the user navigate to the desired content in one or more logical ways. The structure described above is the most typical and addresses all conceivable areas of a typical portfolio setup. Having said that, it is still possible that there may be slight differences in how each organization may want to structure their portfolio website content. Accordingly, there may be small differences in both the main topics as well as the subtopics. However, the main rationale and use of the mind map still hold—it is one of the best tools to show and arrange information in a way that is easy for the users to navigate and understand.

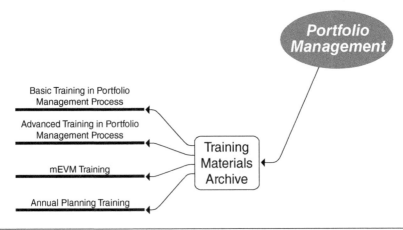

Figure 20.7   Elaborated branch of a mind map for training materials archive

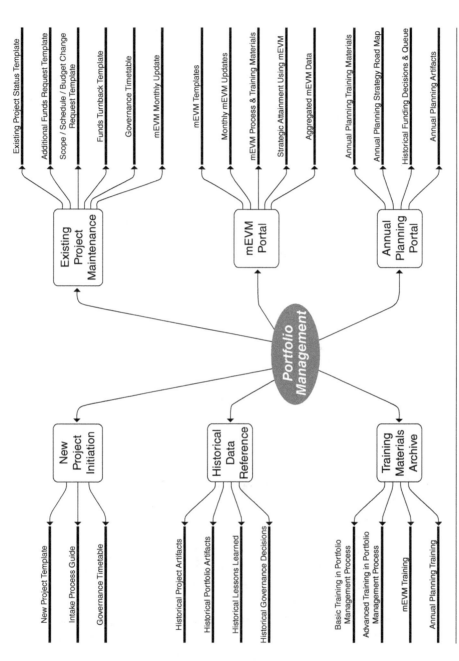

Figure 20.8 Complete mind map with full elaboration of all topics

POINTERS FOR SUCCESS IN USING THE MIND MAP

The mind map is a powerful tool that allows a large mass of information to be effectively navigated and accessed by the partners of the portfolio office. Having a mind map cuts through the clutter of information and lets people get to what they actually want to do—it must always be kept in mind that most stakeholders are only interfacing with the portfolio as a means to an end.

When implemented and managed well, the mind map could prove to be a huge help to the mission of the portfolio office. This is because the mind map provides the user base with a familiar and unchanging interface, although the content behind it may be constantly growing. In this way, a functioning mind map creates a satisfied user base, which is a huge asset for the portfolio office and provides political capital for negotiating additional enhancements to the portfolio's capability. Here are some pointers that enable successful implementation and functioning of the mind map.

- **Pointer #1: Structure the mind map accurately**: For the mind map to serve its function, it needs to be designed in a way that both reflects the hierarchy of content as well as the thinking process of a typical user. The structure narrated in the previous section is a good starting point; however, organizations may have different needs and may have to change the mind map accordingly. The key design principle to keep in mind is that the main topics need to be broad, identifiable anchors and the subtopics need to be progressively more specific and detailed. It also helps to not have too many hierarchical levels in any branch of the mind map. Several iterations of the mind map may need to be performed before a stable version emerges that seems comprehensive and logical.
- **Pointer #2: Validate the mind map initially**: The mind map is meant to serve the needs of the typical user. Accordingly, it is appropriate that before launching the mind map, it should be validated with a group of users to ensure that it works the way it should. For example, the group of users could be given a list of resources that are most frequently searched for and asked to locate these resources using the mind map. User feedback should be carefully considered and changes made as appropriate in response to the feedback.
- **Pointer #3: Train users on how to use the mind map**: Although the mind map is supposed to be intuitively logical to use, it is still a good idea to hold a training session on how to use the mind map. There are a couple of reasons why. For one, most people may not have seen a mind map and may not know how to use one. Secondly, training may serve to

show all of the users that they simply need to start at the mind map in their search for information.

- **Pointer #4: Maintain the mind map regularly**: Although the mind map looks static, it needs to be remembered that it is only an overlay on top of dynamic content. The content changes all the time and the hyperlinks need to change accordingly. Failure to update the links results in broken links, which in turn leads to a loss of confidence on the users. Once users stop trusting/using the mind map, it's very hard to bring them back and the whole artifact becomes obsolete. Therefore, regular updates for the mind map's links and frequent error checking to eliminate broken links are essential to the continued viability of the mind map. At the same time, only authorized personnel in the portfolio office should be tasked with updating the portfolio mind map. Quality control is essential to preserving the integrity and correctness of the mind map.
- **Pointer #5: Leverage the mind map frequently**: A frequently seen and used artifact stays in everyone's mind and continues to be utilized. A less visible artifact is slowly forgotten and ultimately falls into disuse. To ensure that the mind map stays in the limelight, it's a good practice to prominently feature the mind map on the landing page of the portfolio website. It's also a good practice to frequently mention the mind map in portfolio communications, and to include a hyperlink to the mind map in communications e-mails.
- **Pointer #6: Employ mind map software**: Although the mind map described in the previous section was created in Microsoft Visio, mind maps can be built without any software. However, the use of a tool is recommended in some situations. The potential benefits of a mind map software could be ease of use, ability to make changes and publish the output in a common file format such as a PDF, and so on. There are a wide variety of mind map tools available in the market. However, the one thing to keep in mind is that, ultimately, it's the end product (mind map) that matters. If the end product can be facilitated using a mind map software, it may be a good idea that is worth pursuing.

LEVELS OF PORTFOLIO CAPABILITY MATURITY

Level 1

- The portfolio office and resources are likely to be rudimentary with modest capabilities, therefore many changes and rollouts of those changes are needed

- When changes are introduced in the portfolio space, there may not be organized or planned roll-out efforts that accompany changes in the portfolio
- There may not be an adequate estimation of the impact or complete analysis of who is affected by the rollout
- Training may be a *one-size-fits-all* and therefore lack customization
- Training may also lack scenario-based focus and the recipients of training may find it hard to put the concepts learned into practice
- The roll-out effort may not take into account the change management capacity of the organization and hence may attempt to change too much at once
- Training may be imparted, but users may not have the facility to refer to archived training resources at a later time
- No concept of a mind map exists—there may be a website with information that users find hard to navigate due to the mass of information

Level 2

- The portfolio office and resources are likely to be moderately advanced, therefore some changes and rollouts of those changes are still needed
- Some kind of planning and organization exists around roll-out efforts when changes are introduced in the portfolio space
- Prior to roll-out efforts, estimation of the impact and analysis of affected parties may be performed, but this estimation/analysis may not be comprehensive
- Some kind of customization and *tailoring to need* exists for training, but this may not cover all of the different kinds of stakeholders
- Training may include scenario-based actions but all the scenarios may not have been accounted for—consequently, the recipients of training may find it hard to put some of the training concepts into practice
- The roll-out effort takes into account the change management capacity of the organization and hence avoids trying to change too much at once
- Training is delivered by in-house resources whose capabilities may not match that of professional trainers—consequently, training effectiveness may not be optimized
- After the training is delivered, the users may not have the facility to refer to archived training resources, thus depriving them of a vital resource

Level 3

- Although the portfolio office and resources are likely to be well advanced, the constantly improving portfolio office continues to introduce periodic changes and orchestrate the rollouts of those changes
- Roll-out efforts are well planned and organized; sufficient coordination exists to ensure that the roll-out efforts do not clash with other organizational rollouts and/or changes
- Prior to roll-out efforts, comprehensive estimation of impact and analysis of affected parties are performed to ensure that no stakeholder group is unaccounted for
- The needs of the different kinds of stakeholders are taken into account to ensure that training is customized and *tailored to need* as much as possible
- Training prominently features a comprehensive list of scenario-based actions—this ensures that the recipients of training are equipped to deploy the training content in the performance of their day-to-day actions
- The roll-out effort factors in the change management capacity of the organization and hence avoids trying to change too much at once
- Post-training, users have access to archived training resources, thus helping people who may need a refresher about the training content—this also helps users who may not have had a chance to attend training when it was offered
- The concept of a mind map is well known to the organization and forms the gateway to accessing the resources of the portfolio website
- The large and ever-growing portfolio website is navigated without any difficulty by the user base because of the familiar overlay of the mind map, which stays largely constant—and over the years, the addition of significant amounts of data to the website is still scalable and sustainable due to the existence of the mind map

CHAPTER SUMMARY

This chapter explored a key area of portfolio management that is often overlooked—namely, the importance of managing the roll out of portfolio capabilities to the organization. Often, the lack of roll-out management contributes to poor stakeholder reception of what could otherwise be a well-implemented portfolio. The chapter began by reviewing a spectrum of situations that necessitate portfolio roll-out management. This was followed by an analysis of why portfolio roll-out efforts often fail, accompanied by a discussion of the politics surrounding portfolio rollout and change management. The next section in the chapter dealt with recommended best practices for portfolio rollout, which led to an introduction of a powerful technique to organize content—namely the mind map. An extensive treatment of the typical mind map structure for a portfolio office was then provided, followed by a detailed listing of the success factors for a portfolio mind map. The chapter concluded with an analysis of the levels of maturity for this portfolio capability.

21

THE PORTFOLIO OFFICE

INTRODUCTION

The portfolio office is the face of portfolio management and is integral to the success of portfolio management as a function in the organization. In turn, the role of the portfolio manager is central to the portfolio office. In this chapter, we explore the following aspects of the portfolio office:

1. The characteristics of an ideal portfolio manager
2. The strategies in staffing the portfolio office
3. The typical composition of a portfolio office
4. The ideal reporting structure and its place in the organizational hierarchy

DESCRIPTION OF AN IDEAL PORTFOLIO MANAGER

A portfolio manager needs to have a specific blend of skills to be effective. Unlike what some people think, an experienced project manager does not automatically qualify to be a portfolio manager. While the familiarity with projects definitely helps, there is more to being a portfolio manager than that. An experienced program manager, on the other hand, is more equipped to take up portfolio management duties. Why? A program is like a mini portfolio with the added complexity of having the projects in the program be interdependent on each other. Also, a program manager has to contend with financials, intake, and sequencing—all of which find parallels in portfolio management.

Another interesting aspect of the portfolio manager's role is the need to modulate role and tone depending on the audience—the portfolio manager needs to interact with a wide variety of people successfully in order to be effective.

Following are some prime skills/capabilities that a portfolio manager needs to have in their repertoire.

Skill #1: Being Simultaneously Strategic and Tactical

One of the most important skillsets of a portfolio manager is the ability to toggle between the *big picture* and the *fine details*—namely, to see the woods as well as the trees when the occasion calls for it. The portfolio manager needs to grasp the big picture and be able to engage with the people who function at the big picture level—this includes decision makers, strategy formulation experts, as well as Finance, which is concerned with the bottom line. At the same time, the portfolio manager needs to be able to interact effectively with the project managers and program managers, who operate at a far more tactical level.

Skill #2: Being a Process Expert (and Knowing When to Abandon a Process)

Successful portfolio managers are almost always process experts. That's because they recognize the inherent efficiency in a streamlined process, which empowers people to do things the right way in tandem with the rest of the system. At the same time, portfolio managers design simple, lightweight processes that are realistic and workable for their target audiences. They know that process is always (only) a means to an end and should be as simple as possible. Truly outstanding portfolio managers also know when to suspend/abandon process in order to achieve the ultimate goal and are not dogmatic about following process.

Skill #3: Being Data Savvy

A portfolio generates a huge amount of data. The decision makers and other stakeholders are not necessarily interested in plowing through reams of data. It's up to the portfolio manager to read and understand all of the data, as well as being able to sift through the data for insights that then need to be communicated. This requires a fair amount of data savviness—not just being able to read data, but to actually understand what the data is saying.

Skill #4: Thinking Big + Thinking Small

The portfolio manager needs to always have the big picture in mind. Although we covered the need to grasp the strategic big picture, what is being conveyed here is the need to *think big* in all aspects of the portfolio—for example, the need to plan how to take the organization to the next level of portfolio capability. At the same time, the portfolio manager always needs to be in tune with what is feasible and practical on the ground—for example, he or she needs to carefully ponder whether the organization is ready to follow a modified earned value management (*mEVM*) system or ready to undertake annual planning. A successful portfolio manager will always ensure that the little things are taken care of before aspiring to make the big moves.

Skill #5: Being a Credible Operator

While this skill is hard to define, it essentially comes down to being perceived as a credible, result-oriented person by the organization. In other words, the organization needs to have a high degree of confidence in the portfolio manager's competence and reputation in delivering results. This skill is integral to the role because the credibility of the portfolio manager is one of the factors that determines the success of proposed changes in portfolio process.

Skill #6: Being Financially Fluent

A portfolio manager's role is a semi-finance function. Central to the role is the ability to understand how funds are managed within the organization, the timetables of the finance calendar, how the organization classifies operating expenditures and capital expenditures, as well as the overall annual budget cycle and its cadence. In addition to all of this, a portfolio manager needs to interact frequently with the Finance team, and therefore, needs to be able to *speak their language*. Finally, the portfolio manager needs to be seen as knowledgeable in finance processes while conveying trend predictions to decision makers. All of this calls for a high degree of financial fluency.

Skill #7: Being Politically Savvy

Lack of political skills can completely thwart the progress of a portfolio manager who is otherwise skilled in all the aspects of the role. In fact, this one skill can overshadow all of the other attributes—a politically savvy portfolio manager will always find ways to get his or her agenda implemented. They also know when something is too risky to even attempt, and will bide their time to mount the right kind of attack to accomplish their goals. On the other hand, a non-political portfolio manager will frequently find their efforts stonewalled and risk their continued viability in the organization when they try to press their viewpoints in politically adverse situations.

Skill #8: Possessing Project Management Expertise

While we explained that a project manager isn't automatically qualified to be a portfolio manager, it definitely helps a portfolio manager to have good project management skills. A portfolio is high performing only as a result of well-run individual projects. It is impossible for a portfolio to do well with multiple failing projects. Therefore, a portfolio manager has to be equipped with the skills to engage at a project level to understand if the project will fail and weigh down the overall portfolio. The portfolio manager also needs to have good project management skills to engage effectively in rolling out mEVM. Some artifacts like the mEVM excel template are directly related to project execution milestones and the portfolio manager needs to be able to take the lead in creating these mEVM artifacts.

Skill #9: Possessing a Customer Service Mentality

An important skill that is often overlooked while describing a portfolio manager's profile is that the portfolio manager needs to have a strong customer service mentality. Why is this skill so important? Let's consider—a portfolio manager seeks to modify organizational workflow by imposing many process steps and templates. While this is all being done to improve the overall organizational outcomes, it is still an inconvenience in the eyes of the stakeholders. Note that the whole business of portfolio management is seen very differently when viewed from the stakeholders' perspective!

For every change that is rolled out, stakeholders continue to ask many (and often the same) questions over and over again. To accommodate this user behavior pattern and ensure the ultimate success of the effort, the portfolio manager needs to have a customer service mentality to be able to field tedious questions. Having a customer service mentality doesn't just mean having a lot of patience—it involves keeping the stakeholder in mind and designing all changes with a view toward making things easier for the end-user stakeholder.

Skill #10: Possessing Strong Communication Skills

As with all knowledge industry endeavors, communication is a key skill in portfolio management, too. Portfolio managers have the added demand of modulating the same message to suit different audiences. For example, project managers need to be communicated to in specifics; at the same time, decision makers need the same data distilled in a format that lets them make the decisions; and finally, Finance needs an entirely different mode of communication.

Skill #11: Being a Salesman

A successful portfolio manager has to have elements of salesmanship, even showmanship in order to persuade the organization to stay on the portfolio journey. It's common for organizations to feel that they have achieved a lot just by getting to Level 1 capability in most portfolio facets. An organization needs a portfolio manager with good selling skills to convince management to continue investing both political and actual capital in getting to the next level. These skills would also come in handy for the portfolio manager to demonstrate the successful accomplishments of the portfolio office.

STRATEGIES IN STAFFING THE PORTFOLIO OFFICE

In the previous section, we covered all of the skills that a successful portfolio manager needs to possess. First, it's a tall order to find one person who possesses all of these skills. Second, some of these skills are somewhat dichotomous and are unlikely to be found in the same person—for example, the ability to think big while also being detail oriented. Barring rare exceptions, big picture thinkers are typically averse to managing the details and vice versa. It may be more realistic to state that the previous list of skills are capabilities that need to be displayed by the portfolio office. That's why the portfolio office is made up of more than just the portfolio manager. How can the typical organization staff their portfolio office and ensure success? Coming up next are some of the proven strategies.

Strategy #1: Division of Competence

This strategy calls for different people in the portfolio office to play complementary roles, so that as a whole, the portfolio office is able to achieve the desired effectiveness. This works around the need for the portfolio manager to have all of the skills that are outlined in the previous section, which may be an impractical requirement to meet. How could this possibly work?

One way to make this possible is to have a director role (Director of Portfolio Management) and a manager role who report to the director. The manager could conceivably focus on shoring up all of the tactical aspects of the portfolio office and follow ups with project and program managers, while the director could handle all of the senior stakeholders and be responsible for the portfolio office's direction.

Strategy #2: Outsourced Expertise

It's a full-time job to run the portfolio office in its day-to-day functioning and ensure that throughput is kept at an optimal level. However, the portfolio office also needs to continuously improve and get to the next level of capability in all dimensions. How can the portfolio office achieve this *quantum leap* while ensuring the normal operations are not disrupted?

One option is to hire outside help to put in place the arrangements to get to the next level. This may include, for example, consulting to design an mEVM setup, along with the training to roll it out, as well as a website to ensure that people have resources to refer to.

Strategy #3: Political Support and Patronage

Perhaps the most important skill for a portfolio manager is to know how to navigate the political waters of the organization. Without this skill, the portfolio manager cannot prevail in their role and cannot get the portfolio office to where it needs to be. One way of overcoming this hurdle is to entrust the portfolio office role to a seasoned veteran of the organization who is political enough to weather the expected opposition while implementing the portfolio office. This person can then acquire team members who have the required portfolio management skills to actually start and run a portfolio.

Strategy #4: Strength in Numbers

This strategy creates momentum for the portfolio agenda by staffing the portfolio office with numerous personnel who can do much of the work for the stakeholders in an effort to eliminate pushback. For example, consider a situation where the portfolio office fills out the new project templates on behalf of the stakeholders, and then all that the stakeholders have to do is review the template and suggest changes. This would meet with much greater adoption and enthusiasm than, say, asking the stakeholders to fill out the new project proposal completely on their own. Similarly, for most other portfolio actions, the portfolio office can meet the stakeholders more than midway by doing much of the work.

TYPICAL COMPOSITION OF THE PORTFOLIO OFFICE

What should the composition of a portfolio office look like? While it depends on the size of the portfolio, here is a starting point for the kinds of roles to keep in mind while designing the makeup of the portfolio office:

- A director of portfolio management—this should be someone who is typically a senior person with access to executive management. As mentioned before, this person needs to have the political wherewithal to push the agenda of the portfolio management office. This person also needs to be regarded as a credible and reliable partner who keeps their word and delivers results.
- A manager of portfolio management—this requires someone who is well versed in portfolio management and is willing to get involved in the details. This person needs to work closely with the director and provide them with the executive summary (with details as necessary) to convey to executive management.
- Finance role—this needs to be someone who is adept at finance processes and is able to manage the *money aspect* of the portfolio, including interfacing with Finance team members.
- Business analyst—this is someone who is comfortable with pulling reports, rearranging data, producing different views of the data as directed by the manager or director. This person would also own all of the mEVM data and manage the same.
- Project manager—this role is unusual in a portfolio office, but is sometimes seen. This role would provide project management guidelines to the other project managers. This role could also function as the mEVM expert and point of contact for the other project manager.
- Web content management—this role can manage all of the voluminous data being generated by the portfolio operations, maintain the website, and make changes to the website as necessary. This role also needs to be responsible for storing and retrieving historical data for later reference.
- Communications—this would be a role to manage the messaging that comes out of the portfolio office. This person would be responsible for communications regarding the governance schedule, the annual planning timeline, and deadlines.
- Portfolio coordinator—this should be someone who can manage the numerous meetings that need to happen as part of the portfolio operations. This would include scheduling monthly governance meetings, mEVM template update meetings, rollout meetings for new portfolio changes, training sessions for annual planning, and mEVM.

It needs to be kept in mind that this is just a listing of roles/functions that need to be in place for the portfolio to run effectively. More than one role could be done by the same person: for example, the web content management, communications, and portfolio coordination role could be performed by the same person, unless the portfolio is so massive that it makes sense to have a dedicated person filling each of the roles. Also, the director and manager role could be combined, as the situation permits.

PORTFOLIO OFFICE'S REPORTING STRUCTURE

Where should the portfolio office be in the organizational hierarchy? A few broad outlines are provided here, with the understanding that the specifics would vary by organization.

1. The portfolio office needs to have unfettered access to the CIO in order to communicate important decisions that need to be made. See Chapter 18 for more on the close interaction between the portfolio office and the CIO.
2. As mentioned several times, the portfolio needs political power to carry out its role. Where it reports could play a major role in signaling the political clout of the portfolio office to the rest of the organization.
3. From a reporting standpoint, the portfolio office could report directly to the office of the CIO or it could report to a role such as vice president of shared services, who in turn would report to the CIO.

LEVELS OF PORTFOLIO CAPABILITY MATURITY

Level 1

- The concept of a portfolio function is absent—projects are essentially run independently by the respective project managers. A portfolio office, if present, has little power and impact.
- There is little awareness of a portfolio office's value in the organization.
- Sometimes there may be an approximation of the portfolio function, which may only involve one person collecting readouts from all the projects and presenting this status to leadership.
- There is no central management of the portfolio's funds.
- Management does not know how the projects are doing beyond what is self-reported.

Level 2

- The portfolio function exists but is insufficiently defined.
- While there is a portfolio manager, the other supporting roles may not be present, causing the portfolio office to operate at a reduced capacity.
- The portfolio office oversees the execution of projects, but its responsibility is limited due to the basic nature of its capabilities.

- The portfolio office does not have access to executive management and only provides readouts of basic value.
- The portfolio office is not able to carry out funds allocation based on project performance.

Level 3

- A well-defined portfolio function exists with all of the necessary roles that enable the portfolio office to offer a full range of services to the organization.
- The portfolio manager is ably supported by auxiliary roles in the portfolio office.
- The organization recognizes and accepts the role played by the portfolio office.
- The portfolio office oversees the execution of projects in all aspects.
- The portfolio office has access to executive management and provides meaningful recommendations.
- The portfolio office is able to direct resources to the most important strategic activities.

CHAPTER SUMMARY

This chapter covered the team that actually drives the mission of portfolio management forward—namely, the portfolio office. We began by exploring the characteristics of the ideal portfolio manager and the difficulty of finding all of those characteristics in a single person. From that introduction, we went on to describe the strategies that are effective in staffing a portfolio office. We further covered the composition of an ideal portfolio office before detailing the ideal reporting structure and place of the portfolio office in the organizational hierarchy. We concluded the chapter with a description of the levels of maturity for this portfolio capability.

22

THE BUSINESS

INTRODUCTION

Portfolio management is ultimately only a tool for information technology (IT) to accomplish strategic transformation. And IT's prime driver for initiating strategic transformation is to help the business do better in the marketplace. Therefore, it's quite logical to regard the business as the ultimate owner/beneficiary of portfolio management. This chapter will explore the optimal interaction that the business should have with the portfolio office and discuss in detail the following aspects of that relationship:

1. Explore the problematic perception of IT spending from the business partner's perspective
2. List the partnership opportunities that could remediate IT's perception in the eyes of the business
3. List the different ways that the portfolio office could partner with the business during the annual planning exercise
4. Explore how the business could be included in the strategy road map planning exercise
5. Discuss the benefits of including business representation in portfolio governance

A PERCEPTION PROBLEM

The business refers to the part of the organization that is directly involved in the core business of the company. It can be regarded as the portion of the organization that is closer to the market or customer. The business generally views IT as a cost center, and is generally dismayed when allocating budget to IT. "Why are we spending so much on IT?" is a common refrain that is heard during executive budget discussions. In the eyes of the business, IT spends far too much money while not delivering much to show for it. Part of the problem as to why the business does not understand the IT value proposition is that there is (often) no channel for the business to engage with IT in terms of the investments and what they deliver to the strategic transformation.

How can this perception be corrected? Portfolio management can be a bridge between IT and the business in this regard. Many of the portfolio artifacts are valuable in establishing common ground with the business and showcasing IT performance. The following are some strategies that can be used to involve the business and showcase the value of IT through portfolio management, at least so far as project spend is concerned.

Partnership Opportunity #1: Engage the Business in Annual Planning

Annual planning is the process of deciding, as an organization, which programs and projects to fund and execute in the next year. In this process, the whole IT organization is made to articulate all of the collective demand and put it all on the planning table to be prioritized in its entirety. In most places, annual planning is run as a purely IT activity, with IT management *channeling* the viewpoint of the business. Even assuming that IT management is able to serve as a good proxy for the business, this is a one-way arrangement—while IT can understand what the business wants, the business has no visibility to what IT does. This can be remediated by involving the business in annual planning as outlined here:

- **Reflect business priority in annual planning template**: A check box in the annual planning template can be included to show how important the project is to business. This data can then be used while decision makers prioritize demand. (Potential use: compare IT prioritized list to business priority and work through obvious mismatches—for example, what if the lowest ranked IT project shows high business priority?)
- **Include business input in final prioritization of demand**: It's very useful to include the business while creating the final prioritization list because the business then sees and understands what IT is executing on and, more important, how much it costs. There should be much less grumbling about *why we are spending so much on IT* because the business understands exactly where the money is going.
- **Obtain business support to ask for more money**: At some point, the decision makers at the annual planning meeting have to *draw the line*. This represents the point at which the allocated money has run out,

leaving the *below-the-line* projects unfunded. Some of the *below-the-line* projects may be deemed important to the business, which is why IT can obtain business support to show why additional money needs to be allocated to get these projects back *above the line*. This is one more reason why the business needs to be closely involved with annual planning.

Partnership Opportunity #2: Engage the Business in Strategy Road Map Planning

As we saw in numerous preceding chapters, a robust strategy road map forms the essential blueprint for deciding which projects to undertake. It is important to engage the business while developing the IT strategy road map to ensure that all key milestones on the IT road map correspond to valid capabilities on the business road map. This way, when IT delivers on its road map, it becomes possible to demonstrate how it is also enabling the business road map. In order to do that, the following artifact, which was originally introduced in Chapter 11, will need to be modified. Table 22.1 shows the strategy decomposition table referenced as part of the strategic earned value (SEV) concept.

We now add the two additional columns that are listed here, while keeping all of the other columns intact:

1. **Business Strategic Priority**: This column keeps track of what the business strategic priority is. The business strategic plan has its own numbering system (for example BSRP001, which is what is used here).
2. **Business Strategic Priority Description**: This essentially explains what the business priority is trying to accomplish. It is typically less concrete than the IT strategic priority.

The strategic decomposition table would now look as shown in Table 22.2.

This combined table accomplishes the following:

1. **Shows a clear crosswalk between business priorities and IT priorities**: As a result of this table, a clear relationship can be established between business priorities and IT priorities, down to the level of sub-projects. It also establishes the *cost of implementing the road map* and helps answer the question of where IT spends its dollars.
2. **Shows scope gaps between business priorities and IT priorities**: One of the interesting outcomes of completing this table is that it shows where the gaps and nonalignments occur between the business strategic road map and the IT strategic road map. For example, in Table 22.2, there is a business strategic priority *BSRP005* which has no IT equivalent priority and hence, no sub-projects. Essentially, this means that

Table 22.1 Strategy decomposition table showing only IT priorities

IT Strategic Priority	Description	Sub-Projects	Sub-Project Budget	Planned Value	Estimated Completion
1	Real-time payment processing: enable users to pay premiums online	Revamp website to accept online payment (Sub-project 1)	$1 M	$1 M	Year 1
		Implement payment gateway interface (Sub-project 2)	$1M	$1M	Year 1
		Update customer database schema to allow for online access (Sub-project 3)	$1M	$1M	Year 1
2	Self-service for documentation: enable user to print ID cards and proof of insurance through an online portal	Sub-project 4	$1M	$1M	Year 2
		Sub-project 5	$1M	$1M	Year 2
		Sub-project 6	$1M	$1M	Year 2
3	Self-service for claims: enable users to submit claims online including documentation	Sub-project 7	$1M	$1M	Year 3
		Sub-project 8	$1M	$1M	Year 3
		Sub-project 9	$1M	$1M	Year 3
4	Multi-platform capability: enable users to do all the above on a mobile platform in addition to online	Sub-project 10	$1M	$1M	Year 4
		Sub-project 11	$1M	$1M	Year 4
		Sub-project 12	$1M	$1M	Year 4

Table 22.2 Strategy decomposition table showing both IT and business priorities

Business Strategic Priority	Business Strategic Priority Description	IT Strategic Priority	Description	Sub-Projects	Sub-Project Budget	Planned Value	Estimated Completion
BSRP001	Keep parity with competitors who allow real-time payments	1		Revamp website to accept online payment (Sub-project 1)	$1M	$1M	Year 1
			Real-time payment processing: Enable users to pay premiums online	Implement payment gateway interface (Sub-project 2)	$1M	$1M	Year 1
				Update customer database schema to allow for online access (Sub-project 3)	$1M	$1M	Year 1
BSRP002	On-demand access to customer data	2	Self-service for documentation: Enable user to print ID cards and proof of insurance through an online portal	Sub-project 4	$1M	$1M	Year 2
				Sub-project 5	$1M	$1M	Year 2
				Sub-project 6	$1M	$1M	Year 2
BSRP003	Increase customer satisfaction and market share with faster claims processing	3	Self-service for claims: Enable users to submit claims online including documentation	Sub-project 7	$1M	$1M	Year 3
				Sub-project 8	$1M	$1M	Year 3
				Sub-project 9	$1M	$1M	Year 3

				Sub-project 10	$1M	$1M	Year 4
BSRP004	Achieve "most customer friendly" award among all insurance companies	4	Multi-platform capability: Enable users to do all the above on a mobile platform in addition to online	Sub-project 11	$1M	$1M	Year 4
				Sub-project 12	$1M	$1M	Year 4
BSRP005	Become cost leader by reducing premiums of low risk customers	N/A	N/A	N/A	N/A	N/A	N/A

at this point, there are no IT efforts to deliver on a business priority. Unless there are active efforts to address this situation, this may become a serious gap.

3. **Shows schedule gaps between business priorities and IT priorities**: In the table above, imagine that BSRP002 is supposed to be delivered in Year 1. However, the corresponding IT priority and the IT sub-projects are scheduled to be delivered only in Year 2. Unless this is a known issue, this may be a serious gap in expectations.

4. **Establishes a clear setup to demonstrate how IT is delivering on business priorities**: Perhaps the most useful insight derived from this table is the SEV readout. As covered in Chapter 11, SEV can only be recognized when the strategic capability is 100% delivered to the enterprise. For example, the SEV of the project, *Implement payment gateway interface (Sub-project 2)*, is only realized when the payment gateway is implemented and functional. Similarly, Sub-projects 1 and 2 get their SEV only when their respective strategic capabilities are delivered. Here's where this becomes interesting for the business—if all three sub-projects are fully delivered and recognize their SEV, IT management can demonstrate to the business that the IT portion of BSRP001 has been delivered by IT. Some of the advantages that flow from being able to demonstrate such an achievement include:

    a. IT clearly establishes a record of strategic delivery with the business

    b. IT's credibility is clearly underlined with objective evidence

    c. IT is not only able to show that it delivered what was required, but by using the associated modified earned value management measures, it is also able to establish that it delivered within cost and schedule constraints

5. **Enables decision making during portfolio intake**: Matching the incoming project proposals to the business priorities shown in Column A of Table 22.2 will drive a productive discussion of whether a particular proposal that is seeking funding is aligned with the business priority. For example, consider a proposal that is being pushed as a priority by various stakeholders—upon further discussion and analysis, it is found that this proposal does not have a clear alignment with any of the business priorities listed in Column A. This becomes clear grounds for portfolio governance to reject this proposal. Using this artifact regularly ensures that every proposal that is approved for funding is able to be traced back to a business priority.

Partnership Opportunity #3: Include the Business in Portfolio Governance

While the business typically has a dim view of IT spending, IT typically has complaints about the business pushing for ill-advised projects that turn out to yield sub-par benefits. These twin concerns can be solved in the following way: having the business participate in the portfolio governance body can provide both accountability and representation for the business. Representing the business typically involves having a couple of senior business executives join the portfolio governance body. These executives should be able to serve as the voice of the business and be aware of the present and anticipated needs of the business. The accountability stems from having the business executives participate in the portfolio governance process, which includes weighing the pros and cons of all of the incoming proposals, and managing around the constraints of the portfolio. Under this model, the business can no longer claim to be unaware of how and why the portfolio funding and approval process works.

LEVELS OF MATURITY

Level 1

- The organization is not familiar with the concept of business partnership—consequently the business is not regarded as a stakeholder in the portfolio management process.
- The business is unaware of where IT's priorities lie and/or how IT portfolio prioritization works. The business is therefore generally skeptical of IT's funding requests during the budget cycle.
- The business may not be aware of the IT transformation road map (if there is one). If the road map exists and the business is aware of it, there still may be no understanding of how the IT transformation road map aligns with the business transformation road map.
- Business input may not even be included in annual planning and may lead to gaps between business priorities and IT priorities.

Level 2

- An awareness of the concept of business partnership exists, but is not a well-grounded concept—hence, it is not done uniformly or comprehensively.
- Business is aware of where IT's priorities lie in a general sense, but the particulars may be hazy. Therefore, the business is supportive of many of

IT's funding requests, but may be skeptical on some of the IT requests to which there is no visibility of alignment.

- Business may be broadly aware of the IT transformation road map. However, this road map may only be socialized with the business at a high level and there may not be a detailed mapping of business to IT priorities.
- There is a broad understanding of how the IT transformation road map aligns with the business transformation road map, but there isn't an artifact that shows this alignment.
- Business input is included in annual planning at a high level. This may not be comprehensive enough to prevent a few gaps from still existing between business priorities and IT priorities.

Level 3

- A strong awareness exists about the concept of business partnership. This concept is well-grounded and is implemented uniformly and comprehensively.
- Business has been made completely aware of where IT's priorities lie. Therefore, the business is supportive of IT's funding requests as it is aware of the alignment between IT and business priorities.
- Business is well aware of the IT transformation road map. The socialization of the road map occurs not only at a high level but there is also a detailed mapping of business to IT priorities.
- There is an artifact similar to Table 22.2 that explicitly shows the alignment of business and IT transformation road maps. Not only is this artifact frequently used and referred to, there are processes to ensure that this artifact is maintained and updated regularly.
- Business input is included in annual planning at a high level. There exists a comprehensive partnership with the business during annual planning that prevents gaps from existing between business priorities and IT priorities.

CHAPTER SUMMARY

In this chapter, we reviewed the end objective of portfolio management—namely, to enable the business to function better and result in a stronger showing in the marketplace. We then explored a perception problem that exists between IT and the business, which contributes to limited visibility of priorities of the other entity. We then detailed some strategies that can bridge IT and business in the context of portfolio management. As part of these strategies, we introduced an artifact that shows how to map business priorities with IT priorities, and then devolve the IT priorities into concrete projects, thus showing which IT projects align to which business priorities. We also explored how business participation in portfolio governance goes a long way toward bridging IT and the business. We ended the chapter with a look at the levels of portfolio capability maturity.

INDEX